# MARINE ANIMALS
## Partnerships and Other Associations

## Modern Biology Series

*General Editor*

**J. E. Webb, Ph.D., D.Sc.**
*Professor of Zoology,*
*Westfield College, University of London*

---

HORMONES AND EVOLUTION
E. J. W. Barrington, M.A., D.Sc., F.R.S.

SPIDERS AND OTHER ARACHNIDS
Theodore Savory, M.A., F.Z.S.

PLATYHELMINTHES AND PARASITISM
D. R. R. Burt, B.Sc., F.L.S., F.R.S.E.

MICROBIAL AND MOLECULAR GENETICS
J. R. S. Fincham, Ph.D., Sc.D., F.R.S.

ECOLOGY OF FUNGI
R. K. Robinson, M.A., D.Phil.

PHOTOSYNTHESIS
G. E. Fogg, Sc.D., F.R.S.

THE HISTORY OF BRITISH VEGETATION
Winifred Pennington, Ph.D.

WOODLANDS
J. D. Ovington, Ph.D., D.Sc.

# MARINE ANIMALS

## Partnerships and Other Associations

R. V. GOTTO, D.Sc.

*Senior Lecturer, Department of Zoology,*
*Queen's University of Belfast*

*With Illustrations by GLORIA SIDWELL, B.Sc., D.A.(Belf.)*

**THE ENGLISH UNIVERSITIES PRESS LTD**
ST PAUL'S HOUSE · WARWICK LANE · LONDON EC4

ISBN 0 340 04623 6

First printed 1969.   Reprinted 1970

Printed and bound in Great Britain for
The English Universities Press Ltd., by
Richard Clay (The Chaucer Press), Ltd.,
Bungay, Suffolk

# Preface

THE writing of this volume has, in some sense, proved an act of pure self-indulgence, since I have always felt that the subtle gradations which animals reveal in their relations with each other are among the most absorbing phenomena accessible to biological enquiry. Many associations, often involving the most unlikely partners, can be studied on the shore or in coastal waters, and the investigator will quickly find that his observations encompass highly varied aspects of biology—behaviour, evolution, physiology, sensory discrimination and a host of others. It is this very diversity, indeed, which accounts for the subject's compelling charm.

In this necessarily brief survey of marine partnerships, I have tried to present an overall sketch rather than a definitive portrait. Much has been written which deals with the subject at greater length and in a more recondite way. The pressure, however, of current biological courses leaves the student little time to sample these deeper wells of information. I hope that the resulting picture will enable not only university students but sixth-form biologists and the marine naturalist to catch an interesting and profitable glimpse of a relatively little-known world.

R. V. GOTTO

# Contents

# *Contents*

# List of Figures

# List of Figures

# List of Figures

# 1 *Introduction*

ASSOCIATIONS between animals of different species have always held a peculiar fascination for Man. Even in the sophisticated world of today, accounts of cats which foster young rats, or of amicable relations between dog and fox, still rank as newsworthy items. A few years ago the rescue by dolphins of two drowning fishermen off the Japanese coast aroused world-wide, if fleeting attention. Presumably the chief interest of these happenings lies in their seeming reversal of what we have come to regard as a natural law—that the animal world is one of ceaseless, internecine strife. Indeed, one of the most radical changes conceivable to the great prophetic vision of Isaiah was that the wolf should dwell with the lamb.

For this concept of existence as a relentless struggle we must thank not only Darwin, but the great majority of natural historians as far back as Aristotle. It is, after all, true that animals live very much on an eat-or-be-eaten basis. But even the writers of antiquity were aware of certain curious associations involving at the very least toleration, and, at most, clear mutual benefit. Thus Herodotus describes a relationship between the crocodile and a bird—now provisionally identified as the Egyptian plover—which enters the reptile's jaws with impunity in order to feed on the ever-present leeches. And the ancients also knew that, ensconced within the shells of the big dark mussels which they hauled from the sunlit Mediterranean, there lived the small crustaceans which we now call pea-crabs.

A survey of some of these associations and the biological problems which they raise is the theme of this book. Although many animal partnerships can be found on land and in fresh water, our examples will be drawn almost exclusively from the sea. There are several reasons for restricting our enquiry to the ocean. First, a truly comprehensive study embracing all known forms of association is beyond the scope of this small volume. Secondly, marine biological courses are now so frequently recognized as part of University curricula that interest in marine life is widespread. Finally, and perhaps most compellingly, the sea affords unique opportunities for studying this particular aspect of biology. Not only is it spatially the greatest of Earth's environments, but it is also the oldest. Here the long magic of evolution has worked undisturbed to produce an incredibly diversified array of animal types. Cradled by the dense watery medium, patterns of organic architecture, un-thinkable in other environments, can here take shape. And this great struct-ural diversity is paralleled by habits and ways of life which are often bizarre to the point of fantasy. Against such a background, it is not surprising that the whole spectrum of animal associations, from the obviously casual to the intimately complex, can be seen.

Since many partners show complementary and intricate adaptations to a shared mode of life, it is clear that some of these associations must be very old indeed. However, so many degrees of intimacy are apparent that it is equally obvious that some are likely to be of comparatively recent origin, and it should be remembered that even a slight shift in behavioural pattern may be sufficient to trigger a new type of association. To take a single example, the tits which industriously peck through the metal caps of milk bottles, have forged a new link in their relationship with Man over the past few decades.

Many different terms have been used in an attempt to define and describe animal partnerships. The complex relations existing between the social insects and their characteristic associates alone account for a considerable number. A fair proportion, however, are used in less specialized contexts, and have been widely applied in association studies. Unfortunately, authors find the greatest difficulty in using these terms in a precisely similar manner, and the same association will often be described under different headings according to the authority consulted. There are three perfectly good reasons for this. In the first place, as all writers agree, our knowledge of the total biological picture is often only fragmentary. Secondly, one particular feature of the association may strike an author as having paramount significance, and thus influence his choice of definition for the relationship as a whole. Thirdly, Nature reveals such subtle gradations that the imposition of hard and fast boundaries is well nigh impossible.

Now let us examine some of these terms, remembering always that they are arbitrary, and decide the sense in which we are going to employ them through-out this book.

*Endoecism.* This describes a partnership in which one animal habitually shelters within the tube or burrow of another. Shelter or protection is thus the dominant factor, though in some cases an endoekete may also have a food-sharing (commensalistic) relationship with its host.

*Inquilinism.* This was originally used to define a nest- or gall-sharing association among insects, but has since acquired a somewhat different significance. Here we shall employ the term to denote organisms which live together, one within the other, the former utilizing the host animal mainly as a refuge. Once again, therefore, protection is the keynote, but these associations imply a rather greater degree of intimacy between the partners than is found in endoecism.

*Phoresis (Phoresy).* This arises from the Greek word *pherein* meaning "to bear or carry", and was first used to describe the transport of various small arthropods by larger insects. In recent times the definition has been broadened, and some authors now use this term where others would employ inquilinism. However, the original concept of "transport" (or even the host's movement) as an important feature in such partnerships, is a useful one and should be preserved. We shall accordingly define phoresis as indicating organisms which live together in such a way that the transport provided by one promotes the well-being of the other.

*Epizoism.* An epizoite can be defined as an animal which lives (attached or free) on the surface of another animal. This may involve a rather marked substrate preference and/or the provision by the host of a precise set of favourable environmental conditions, though without implying ectoparasitism.

*Mutualism.* As the word itself suggests, mutualism denotes an association involving reciprocal benefit. It is often used as a synonym of symbiosis—a term which has had a particularly confusing history. Here, however, we shall restrict it to cover associations in which mutual advantage is apparent, but without the implication that the partners are ultimately dependent on one another in a physiological sense. The cleaning associations described in Chapter 5 constitute a specialized form of mutualism.

*Commensalism.* The word commensalism means "at table together" and in classical usage denotes organisms which live together, with no harm to either, and which generally share a source of food. The actual process of obtaining the food, however, is usually carried out by one of the partners only. The advantages are thus one-sided, the commensal being, in effect, a non-paying guest. Commensals which live on the external surface of the host are called ectocommensals, while those which occur inside the body are referred to as endocommensals.

*Symbiosis.* This term was used by De Bary in 1879 in its literal and very broad sense of "living together". As such, it included not only co-operative partnerships but commensalism and parasitism as well. Although of late years there has been a tendency to revert to this original wide definition, one cannot disguise the fact that for the majority of biologists the term implies a partnership of mutual benefit. Some modern authors, in fact, have restricted its use to cases in which the associates have come to rely on one another physiologically, so that separate existence is impossible. Here perhaps we can steer a middle course and define symbiosis as a state of affairs in which organisms live together for mutual benefit in such close relationship that each has lost at least some degree of physiological independence.

Two examples should make this clear. Both, as it happens, are terrestrial, but their illustrative value is considerable, since they have been the subject of many experiments on symbiosis as defined in this way.

Many primitive termites (*Calotermes* for instance) contain a thriving fauna of specialized flagellate protozoans, such as *Trichonympha*, which live exclusively in the hind-gut. The termites eat wood, but lack the cellulose-splitting enzyme which their flagellates possess. The latter, therefore, play a vital part in the termites' digestive process, breaking down the cellulose into various products such as glucose and acetic acid, which thus become available to the insect host. The termites further benefit by ultimately digesting many of their unicellular partners, which represent a significant source of protein. On the other hand, the flagellates are intolerant of oxygen and so find ideal living conditions in the termites' gut, where the oxygen tension is very low.

This is a complete symbiosis, since neither termite nor flagellate can survive alone. A somewhat similar association exists between various ciliate protozoans and ruminating mammals such as cattle. In the rumen, cellulose is again broken down by certain of the ciliates and a surprising quantity of protein (about 20 per cent of a ruminant's requirements in fact) made available to the host when the micro-organisms are finally digested. In this instance, the warm, methane-filled paunch is the only environment in which the ciliates can live and thrive. However, physiological dependence is not so marked in this case, since the ruminant can do without its protozoan fauna so long as cellulose-splitting bacteria are present. Other examples from the marine environment, probably involving equally close metabolic interplay between partners, are given in Chapter 7.

Now it is obvious that a dividing line between the definitions adopted for mutualism and symbiosis will be fine to the point of virtual invisibility. In the present state of our knowledge it is often very difficult to assert that associates are *not* in a physiological equilibrium so subtle as to defy detection by current methods. One could perhaps cut this particular Gordian knot by referring to symbioses of *proved* physiological interdependence as "obligatory mutualism", and to less intimate, but still reciprocal, partnerships as "facultative mutualism". For the moment, however, such hair-splitting must remain a matter of personal choice.

*Parasitism.* This is a word for which a really satisfactory definition is still awaited. It will perhaps suffice to take the line of least resistance and present it in the generally understood sense—that is, the state in which organisms live together such that advantage accrues only to one partner, while the other is frequently harmed or debilitated. This admittedly begs many questions, but parasitism as such will not figure largely in this book, which makes no pretence of being a treatise on this aspect of zoology. Many of the features and adaptations of truly parasitic forms are common to all environments and will have been adequately studied in general zoological courses.

It must be emphasized again that the associations covered by these definitions can very easily grade into each other. In most cases a great deal more information is needed before a partnership can be satisfactorily classified. At present the onus is very much on the individual worker, his choice of category depending largely on what aspect he elects to stress—shelter, transport, food-sharing and so on. Some authors have, to a certain extent, overcome this difficulty by referring to only three types of shared existence—symbiosis, commensalism and parasitism. Thus any partnership which is definitely neither symbiotic nor parasitic comes under the heading of commensalism. The latter term is in this way converted into a sort of lumber-room into which are thrust partnerships of widely different kinds, the original and useful connotation of a shared food-source often being abandoned in the process. While there may be something to be said for this broad and simple approach, the existence and current use of the other available categories should be recognized. The real answer is for more intensive investigation into the exact nature of these often fascinating associations.

# 2   *The Refugees: Protective Associations*

U NDER this heading we can discuss various associations which span a number of the categories mentioned in Chapter 1. Several of them, for example, combine the connotation of food-sharing (true commensalism) with that of shelter (inquilinism and endoecism). In others, again, some degree of reciprocal benefit (mutualism) can be discerned. This merely serves to emphasize how difficult it is to lay down hard and fast rules regarding types of partnership. In the main, however, we shall concern ourselves with those associations in which the provision by a host of a refuge of some kind would appear to be the over-riding factor in the business of living together.

## Protection by proximity

Probably in few other habitats are animals so exposed and vulnerable to attack as in the surface waters of the open sea. Here, surely, is no hiding place! The small and feeble in this bright and naked world of water may be struck down at any moment in a score of different ways. It is true that many, like the arrow-worms, gain protection from a ghost-like transparency which renders them virtually invisible, but for the others survival must be very much a matter of pure chance. Yet even in this hostile place, refuges exist, scattered like oases in the desert of the sea, available for those fortunate enough to find them and sufficiently adaptable to use them.

These havens are provided by the large pelagic jellyfish which are common oceanic forms. Almost any scyphozoan of reasonable size is liable to shelter a variety of small animals beneath its slowly pulsing umbrella. Young fish are often found in this situation, and belong to a surprisingly wide range of species. Horse mackerel, haddock, whiting, cod, yellow-tails and amberjacks

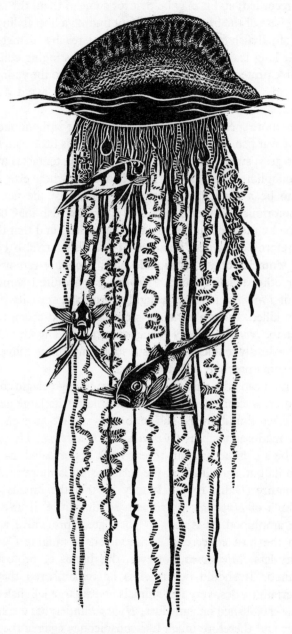

**Figure 1** The siphonophore *Physalia* with its attendant fish, *Nomeus gronovii*. (Adapted from various sources.)

have all been recorded, and it may be that for some of them the association is an obligatory phase of the life cycle, those not finding a host failing to survive. Some species of jellyfish (e.g., *Cyanea arctica*) may reach a diameter of six feet and, with their long trailing tentacles bristling with stinging cells, it is clear that considerable protection from predators is afforded to the young fish which accompany them. Just how these juveniles themselves avoid destruction is still problematical.

A similar but more specific partnership is that of the siphonophore *Physalia* and the man-of-war fish *Nomeus gronovii* (Fig. 1). This little fish, striped with blue and silver-grey, swims among the immensely long tentacles which dangle from the siphonophore's float, and is in fact found nowhere else. One would imagine this to be a particularly perilous niche, since the nematocysts of *Physalia* are notoriously virulent. It is said that other fish may be lured into the danger zone by the presence of this small associate, and that the resulting meal may therefore be shared. The man-of-war fish has also been seen to nibble at the tentacles of its formidable host, probably removing trapped crustaceans or other small animals. *Nomeus* has acquired some immunity to the toxins of *Physalia*, since, if roughly thrown against its partner, it is not necessarily killed, though certainly stung. It would seem, then, that in the natural state, *Nomeus* simply manages to avoid any but the slightest contact with the deadly tentacles. If this is so, its swimming movements must surely involve some precisely adapted reflexes.

Fish are by no means the only animals to accompany pelagic coelenterates. The amphipod crustacean *Hyperia galba*, notable for its large and beautiful green eyes, is often found on the sub-umbrellar surface or even in the sub-genital pits of medusae such as *Cyanea* and *Rhizostoma*. *Hyperia* is often considered to be a commensal, but recent work has revealed the presence of nematocysts in its gut, showing that the relationship is more probably of an ectoparasitic nature (Dahl, 1959). It may be that the mucus of the host provides the bulk of *Hyperia*'s nutriment, so that little, if any, damage is caused by the amphipod. It is easy to postulate a protective advantage to *Hyperia* when the host concerned is the viciously stinging *Cyanea*. With the almost harmless *Rhizostoma*, however, this bonus of protection is presumably forfeited. It should nevertheless be remembered that, with the exception of certain turtles, very few animals attack large jellyfish of any sort. Moreover, once established on a medusa, *Hyperia* undergoes a colour change, becoming paler and therefore much less conspicuous against the translucent tissue of its host.

Protective associations involving fish and echinoids are also well established. The warm-water sea-urchins of the genus *Diadema* possess long, fragile,

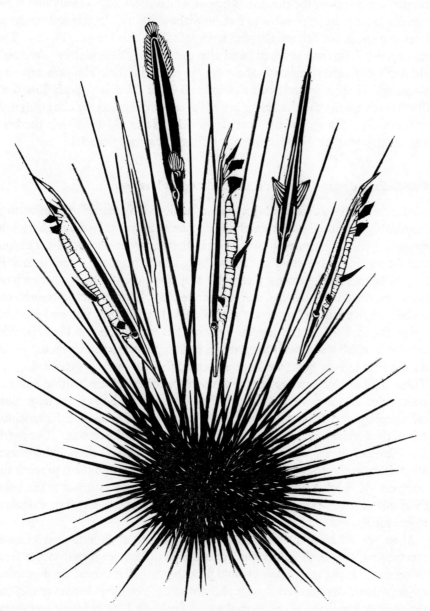

**Figure 2**  Three shrimp-fish (*Aeoliscus strigatus*) and two cling-fish (*Diademichthys deversor*) shelter head-down among the spines of the sea-urchin *Diadema*. (Adapted from a drawing by N. B. Marshall.)

needle-sharp spines, the poisonous tips of which break off so easily that these urchins are almost impossible to handle without injury. In tropical waters at least two species of fish are adapted to benefit from this feature (Fig. 2). The shrimp-fish (*Aeoliscus strigatus*) and the cling-fish (*Diademichthys deversor*), although not closely related, show convergence in their elongate tapering shape and in their specialized behaviour *vis-à-vis Diadema*. Both float head downwards among the slender spines of the sea-urchin, and the cling-fish may at times nip off some of the host's tube-feet. It would be a rash predator which attempted to dislodge these fish from their spiny citadel.

## Pearl-fish, damsel-fish and specialized behaviour

So far, we have considered protection achieved by mere proximity to a formidably armoured host, though the proximity is maintained by behavioural means. A more intimate partnership is that between the pearl-fishes and various marine invertebrates. These fish, which measure a few inches in length, belong to the family Carapidae and upwards of twenty species have been described from tropical and temperate seas. Their scaleless translucent bodies are elongate and narrow, the pelvic fins have disappeared and the anus opens far forward in the jugular region. Several species of bivalves, sea-urchins, starfish and sea-squirts have all been reported to shelter them, but by far the best known hosts are large sea-cucumbers of various genera (Fig. 3). These big holothurians are very prominent members of the benthic fauna in warm seas. At the posterior end of the sausage-shaped body is the large cloacal opening, and from the cloaca itself a system of thin-walled branching tubes, the respiratory trees, extends well forward into the coelom. The pearl-fish (*Carapus*) gains access to its host through the cloacal pore, living thereafter either in the respiratory trees or else breaking through them to reach the body cavity. Such a rupture is probably of no great consequence to the holothurian, since it possesses the usual echinoderm capacity for extensive regeneration.

Most pearl-fish seem to venture outside their host at night to feed on small crustaceans, and both adults and juveniles have been captured in the free-living state. However, *Carapus* would probably fare ill if denied the protection of its partner. Its first larval stage, known as the vexillifer, is planktonic, but the second (tenuis stage) is dependent on finding a host before it can metamorphose to the juvenile condition.

The whole of this association strongly suggests continuous adaptation over a long period of time. In itself, the behaviour of the fish indicates specialized responses to its peculiar mode of life. Thus it can be attracted to a model only if a current of water containing holothurian mucus is passed through it (Arnold, 1957). Although young fish will sometimes enter the host's body head-first, the adults have a different method. After pushing the snout against the cloacal opening they rapidly insert the tail and so work their way in backwards. Sometimes the cucumber, perhaps not unnaturally, attempts to close its vent against the intruder, in which case the fish adopts a screwing action and gradually forces itself in.

**Figure 3**   The pearl-fish *Carapus* entering its holothurian host tail-first via the cloacal pore. (Adapted from various sources.)

The structural modifications of *Carapus* likewise indicate that this is an association of long standing. The loss of scales and pelvic fins and the forward shift of the anus, which ensures the voiding of faecal matter outside the cucumber's body, are strikingly adaptive features. In choosing a holothurian, moreover, the pearl-fish has selected a host which is rarely disturbed by predators. This comparative immunity from attack may be due in part to the extremely toxic substances frequently present in holothurian skin.

The pearl-fish partners of sea-cucumbers thus rank as inquilines. We should note, however, that the Mediterranean pearl-fish, *Carapus acus*, is known to browse on the gonads of its host, so that in this species the relationship may be grading into parasitism.

Some of the cardinal fishes (e.g., *Apogonichthys stellatus* and *A. puncticulatus*) have what appears to be an association of somewhat similar type with large conch shells of the genus *Strombus*. They find shelter in the mantle cavity and, like *Carapus*, emerge from their host mainly during the hours of darkness.

A case of special interest is that of the small, brightly coloured damsel-fish

**Figure 4**   Damsel-fish with their host anemone. (Adapted from various sources.)

belonging to the genera *Amphiprion* and *Premnas*, which team up with giant sea-anemones on tropical reefs (Fig. 4). Often a pair of damsel-fish will adopt a particular anemone, which thereafter acquires a territorial significance. They swim unharmed among the tentacles of their host, although other fish of similar size are quickly seized and eaten. *Amphiprion* can even be trapped for short periods within the anemone's gastral cavity without suffering any apparent inconvenience.

Recent work on *Amphiprion percula* and its host *Stoichactis* has shown that the success of this association depends on a series of minutely adjusted behavioural and physiological responses (Davenport & Norris, 1958). Visual recognition of the host is the first step. There follows an "acclimation" process, in which the fish swims closer and closer to the anemone, using a characteristic series of slow vertical undulations. Ultimately a brief contact is made, generally causing a clinging reaction of the tentacles, from which the fish frees itself with a violent jerk. With increasing regularity of contact, the clinging reaction diminishes and finally ceases. *Amphiprion* now swims among the tentacles with added speed, evoking no response from *Stoichactis* although its tentacles are by this time in almost continual contact with the fish's skin. This acclimation process requires on the average about one hour, and that it is a very necessary ritual is shown by the fact that a non-acclimated damsel-fish is liable to be seized and devoured. Experiments have revealed the presence of an active principle in the mucus secreted by the skin of *A. percula*, which raises the threshold of mechanically-induced discharge of the anemone's nematocysts. This factor is fast-acting, specific in its effect and heat-labile. It has furthermore been demonstrated that a similar immunity characterizes the eggs of *Amphiprion*, which completely fail to evoke a clinging reaction from the host's tentacles.

Are we confronted here with an instance of preadaptation—the fortuitous possession by damsel-fish of a mucus factor inhibiting nematocyst discharge in certain large anemones? Or are we to visualize this partnership as the climax of a long series of cumulatively selective processes? Whatever its origin, there can be little doubt that the association as we see it today reveals an amazing degree of precision and refinement.

What of the relationship between the two? *Amphiprion* clearly gains protection from its large partner. Under aquarium conditions it is generally eaten by other fish if isolated from its host. If advantages accrue to the latter, they are not so obvious. There is, however, some reason to believe that damsel-fish fulfil a scavenging role, eating the digestive wastes of the anemone. The suggestion has also been made that they may act as decoys, and further that they may serve to keep down parasitic growths and infestations on the anemone's skin. One species, *Amphiprion polymnus*, has been observed to bring pieces of food to its host and place them among the tentacles, but whether this happens often in nature is still doubtful. Nevertheless the association may well be of mutualistic type.

## The protective role of host burrows

A partnership of a much looser and less specific kind, but one in which shelter is still the dominant factor, can be found between the arrow-goby *Clevelandia ios* and various burrowing invertebrates along the North American coast (MacGinitie & MacGinitie, 1949). These small, cryptically coloured fish about $1\frac{1}{2}$ inches in length, are particularly common in sandy or muddy estuaries. At the approach of danger, and when the tide goes out, they tend to seek refuge in any hole available. The burrows of the large echiuroid worm *Urechis caupo* and those of the burrowing shrimps *Callianassa* and *Upogebia* seem to be particularly attractive (Fig. 5). Indeed, up to two dozen gobies have been recorded from a single *Urechis* burrow. Although they range freely in search of food, their utilization of other animals' homes is a very real and characteristic feature. Moreover, *C. ios* is nothing if not an opportunist. When the burrow is shared with a pea-crab, it has been observed to bring a piece of food too large to be swallowed to the crustacean, which obligingly shreds the morsel into more manageable fragments. One interesting aspect of the arrow-goby's intrusion is the lack of agitation shown by the host. This is clearly a well established case of facultative endoecism, and the toleration shown to *Clevelandia* indicates that the relationship could perhaps in time become closer and more permanent.

On the same coast, *Callianassa affinis* acts as host to another little fish, the blind goby, *Typhlogobius californiensis*. When young, these gobies can swim actively and have normally developed eyes, but after some months they take up permanent residence in the burrow of a ghost-shrimp, a male and female generally being found together. By this time the skin has paled to a flesh colour and a membranous growth covers the eyes. The host's burrow now constitutes a territory, from which a rival male will be driven vigorously, although *Typhlogobius* is usually rather sluggish. Its general pattern of inactivity, in fact, enables it to survive in almost stagnant water.

This association differs from that of *Callianassa-Clevelandia* in that the blind goby is an obligatory endoekete. However, the partnership is not, strictly speaking, on a food-sharing basis, since different food-sources are involved. In their burrows, ghost-shrimps dig for a living and sift minute particles of detritus from the disturbed mud. The goby, on the other hand, feeds partly on small animals, but mainly on seaweed debris which collects in the burrow. It may therefore play some small part in maintaining the communal shelter in a state of cleanliness. The shrimps, which are industrious and fastidious house-keepers, will sometimes collect small pieces of weed and deposit them in front of the gobies. On the whole, however, *C. affinis* thrives perfectly well without its lodger.

A clearly established instance of fish and crustaceans finding each other's company beneficial is known from the Red Sea. Here pistol-shrimps of the genus *Alpheus* tend to form a partnership with two small gobiid fish, *Cryptocentrus caeruleopunctatus* and *Vanderhorstia delagoae*. These fish prefer living in the safety of holes or burrows, at the entrance of which they can normally be found, and the excavations made by *Alpheus* are especially favoured. As long as the fish is standing guard, *Alpheus* continues to burrow quite happily,

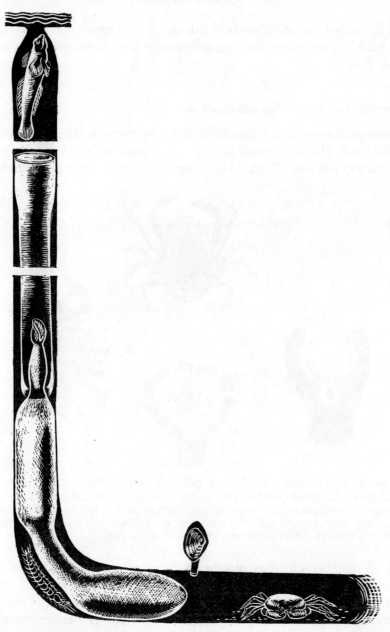

**Figure 5** *Urechis caupo* and some of its associates. Here the burrow is shared with an arrow-goby (*Clevelandia ios*), a pea-crab (*Scleroplax granulata*), a scale-worm (*Hesperonoë adventor*) and a clam (*Cryptomya californica*). Note the tubular feeding-funnel of slime secreted by *Urechis*. (Adapted from Natural History of Marine Animals, by MacGinitie and MacGinitie, by permission of the McGraw-Hill Book Company.)

but at the first sign of danger both fish and shrimp retreat into the hole, and the latter will not resume its digging operations until its piscine sentry is once more posted.

## A variety of protective associations

A very good example of protection is shown by the coral gall crabs belonging to the family *Hapalocarcinidae* (Fig. 6). The female settles down at the growing point of a branch of coral. Her presence stimulates the coral to produce a

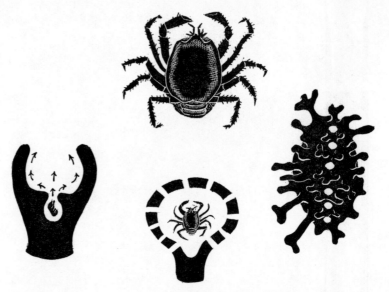

**Figure 6**  Gall formation by the coral gall crab *Hapalocarcinus*. *Top*, adult female. *Bottom left*, currents produced by the newly-settled young crab cause the growing coral branches to arch. *Bottom centre*, diagrammatic section of a completed gall. *Bottom right*, mature gall with sprouting coral shoots and respiratory apertures (white circles). (Adapted from a drawing by F. A. Potts.)

gall or chamber which eventually encloses her, though a series of small apertures remains which keep the crab in communication with the outside world. Through these holes *Hapalocarcinus* causes a current of water to circulate, from which it filters minute food. It is also via these entrances that the tiny males gain access to the now permanently immured female for the purpose of fertilization. In an analogous manner, a large gall may be formed on the apical region of the sea-urchin *Echinothrix turca*, due to the presence of the crab *Eumedon convictor*, which habitually settles in the urchin's rectal pouch.

In tropical seas a curious alliance has evolved between the sipunculid worm *Aspidosiphon* and the solitary coral *Heteropsammia cochlea*. *Aspidosiphon* lives in the small, empty shell of a gastropod. The larval polyp establishes itself on the shell and presently overgrows it. The sipunculid responds by maintaining a gallery within the basal portion of the polyp, from which it can protrude and move about, dragging along its passive partner. Protection is thus afforded to *Aspidosiphon*, while the coral may possibly benefit by being transported to fresh feeding grounds. A similar advantage of free carriage to new pastures may be gained by hydroids of the genus *Hydractinia*, which regularly colonize shells inhabited by hermit crabs. Since the spiral zooids of these colonies are armed with nematocysts, the hermits are assured of a safe refuge.

Sponges occur frequently as participants in marine associations. Thus

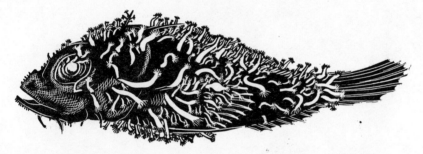

**Figure 7** The scorpaenid fish *Minous inermis* largely covered by the hydroid *Podocorella minoi*. (Adapted from a drawing by T. Komai.)

some hermits may be almost engulfed by the exuberant growth of *Suberites*. This same genus may again be found in a more specialized partnership with the dromiid crabs, which deliberately "plant" their associate, holding the sponge with their clawed hind legs until it has become fixed to the dorsal surface of the carapace. In this case the benefit would seem to be purely one-sided, the spicular nature of sponge skeletons rendering them distasteful or inedible to all except a few specialized predators.

The safeguarding of the vulnerable egg stage is probably the *raison d'être* for a peculiar association between the liparid fish *Careproctus sinensis* and certain crabs. The female fish is provided with a long ovipositor with which she affixes her eggs below the crab's claws. A better guarded nursery than this would be difficult to conceive.

An indirect type of protection in the form of effective camouflage is, of course, operative when epizoic organisms settle in significant profusion on some other animal. Sometimes these associations are of a highly specific nature. For example, the scorpaenid fish *Minous inermis* from Indo-Pacific waters has never been found without a dense coating of the hydroid *Podocorella* (= *Stylactis*) *minoi*, nor have these polyps been observed in any other situation (Fig. 7). This argues that fish and hydroid interact ecologically in a narrowly specialized manner.

Enough has been said to show that, for many animals, refuge and consequent survival do not necessarily depend on the crevices, holes and other shelters provided by the inorganic environment. Protection is often guaranteed by the complex and delicately balanced relationships which have evolved between widely different members of the marine fauna.

# 3 *The Hitch-hikers: Transport Associations*

THE concept of transport as a feature of importance in certain animal alliances will now be reviewed briefly. This idea, however, is not meant to imply that *dispersal* through the agency of a host is the dominant consideration, although obviously a wider distribution frequently results from an association of this kind. We are concerned here rather with the immediate consequences of a phoretic partnership on the day-to-day life processes of an individual animal.

## Specificity of host movement

Phoresis in which this is an important factor has been noted more often in fresh-water than in the sea. Here many specialized ciliate protozoans live attached to the surface of aquatic insects and crustaceans. They are not in any sense parasitic, but rely on the movement of the host to set up currents from which they can pluck their food. Several species of the peritrichan *Opercularia* are fixed in this way to a variety of aquatic invertebrates. In such cases, we may find a marked degree of specialization, in that the ciliates are intolerant of transplantation to a different host. It is evident that their feeding habits are closely adjusted to a specific pattern of host movement.

Perhaps due to the near-absence of insects in the sea, analogous examples are harder to find in the marine environment. However, the suctorian protozoan *Dendrosomides paguri* occurs on the setae adorning the limbs of hermit-crabs, and we may assume that the movements of the pagurid are an important factor in the feeding activities of its little partner. In the same way, a species of the peritrichan *Rhabdostyla* is found in the Irish Sea on the gills and body of the terebellid worm *Nicolea*, and here again the host's movement is probably vital for the well-being of the ciliate.

## Specialized barnacles and their adaptations

The barnacles which are characteristically attached to whales are of particular interest in a phoretic context. A fair number of whale-barnacles are known, and this highly specialized niche has been independently exploited several times during the course of cirripede evolution. This is evidenced by the fact that the barnacles of Cetacea belong to two distinct families. Thus *Tubicinella* and *Coronula* are of balanid type, like the familiar acorn-barnacles of the littoral zone, while *Conchoderma* (Fig. 8) is a lepadid, and so resembles the goose-barnacles sometimes washed ashore on pieces of driftwood. *Xenobalanus* is an extraordinary form in that it is a balanid which has convergently acquired a startling similarity to the lepadid *Conchoderma*.

So many barnacle species are attached to hard, non-living surfaces, that it is interesting to study the adaptations imposed by life on a relatively soft organic substrate. *Tubicinella*, for example, has transverse ridges on its shell-valves which grow into the host's epidermis to such effect that the barnacle appears to be screwed into the whale's skin. *Coronula* achieves a similarly firm anchorage, and indeed the removal of these epizoites leaves quite noticeable scars and craters. Moreover, since recent work has emphasized the extraordinarily selective behaviour of cirripede larvae in their choice of settlement, we must suppose that evolution has impinged with equal effect on the larval stages of whale-barnacles and moulded their response to surface texture in a very specialized way. We shall discuss this point a little more fully in the next chapter.

The genus *Conchoderma* seems to have retained a greater degree of plasticity in this matter of larval settlement. *C. auritum*, although capable of colonizing whale skin, is usually found attached to the shell-valves of *Coronula*. It has also been recorded from the baleen plates of whale-bone whales, and even from the teeth of a sperm whale. In this latter case, however, the whale concerned had a pronounced curvature of the mandible, so that the teeth did not fit into the groove of the upper jaw. It was presumably this deformity which permitted settlement in so unusual a site. Another host of *C. auritum* is the large eel *Gymnothorax*, while the sides and bottoms of ships have furnished additional records. An allied species, *C. virgatum*, has been found on driftwood, ships' bottoms, the sunfish *Mola*, the eel *Gymnothorax*, sea-snakes, a decapod crustacean and the large copepod *Pennella* which itself is embedded in the skin of whales and marlins.

**Figure 8** Three specimens of the whale-barnacle *Conchoderma auritum* fixed to another whale-barnacle, *Coronula diadema*. Note the rabbit-like "ears" of *Conchoderma* through which water leaves the barnacle's hood. (Adapted from Natural History of Marine Animals, by MacGinitie and MacGinitie, by permission of the McGraw-Hill Book Company.)

B

*Conchoderma auritum* is a soft-bodied form, the shell-valves having virtually disappeared. The feathery, food-trapping legs or cirri are protected and enclosed by a sort of hood, the open side of which is generally orientated towards the whale's head. Two funnel-like tubes extend back from the hood, their resemblance to ears giving this cirripede the popular name of "rabbit-eared barnacle". As the whale swims, the water current flows through the hood's forwardly directed opening. Planktonic organisms are then trapped by the cirri, while the water passes out via the two "ears".

Whales are not the only vehicles available to specialized barnacles. *Chelonibia* attaches itself to turtles, firm connection being ensured by a branched root-like system which penetrates the bone of the plastron. It has also been recorded from manatees or sea-cows. *Platylepas ophiophilus* is associated with the sea-snake *Enhydris curta* in Malaysian waters, while some members of the genus *Alepas* occur on medusae. In all these cases it may be assumed that the host's movement is of functional significance to the feeding activities of its cirripede partner.

These subtle accommodations to currents induced by the host are beautifully illustrated by the tiny stalked barnacle *Octolasmis stella*. This is found on the gills of the lobster *Puerulus sewelli* in the Indian Ocean, each individual being orientated with relation to the respiratory flow over the gills and through the branchial chamber generally. The final, fixed position of the adult appears to depend on the precise alignment achieved by the cyprid larva at settlement (Dinamani, 1964).

### Sucker-fish and their hosts

No account of marine phoretic associations would be complete without mention of the so-called sucker-fish or remoras. These belong to the family Echeneidae, and their remarkable powers of adhesion have been known since very ancient times. About ten species are recognized, mainly from tropical waters. Remoras range from a few inches to three feet in length and attach themselves to a variety of moving objects, several individuals often occurring on a single host. Attachment is effected by means of a large oval sucking disc on the top of the fish's head (Fig. 9). This disc, which represents a highly modified anterior dorsal fin, has a slightly elevated rim and is crossed by a series of transverse ridges, which are probably transmuted fin-rays. Since the raising and lowering of the rim is under muscular control, suction can be created when the disc is brought into contact with the host's skin. In this way, remoras are towed passively by their large partner. When the latter obtains food, the echeneid releases the suction, slips its moorings and swims forward to share the meal. *Echeneis naucrates* accompanies sharks, giant rays, sting rays and ocean sun-fish; *Rhombochirus osteochir* is generally found with marlins and sail-fish, while *R. brachypterus* associates with marlins, tunny and swordfish. *Phtheirichthys lineatus* has been observed on sharks, tarpon and barracuda, and *Remiglia australis* on whales. Turtles too may be utilized, and echeneids will also attach themselves to ships. In some parts of the world, captive remoras are used to catch fish and turtles, a line being woven around the shark-sucker's tail, and in Madagascar these fish have their place in sorcery. A dried portion of the disc is hung around the neck of an unfaithful wife, presumably in the hope that its well-known adherent properties will stabilize a crumbling relationship!

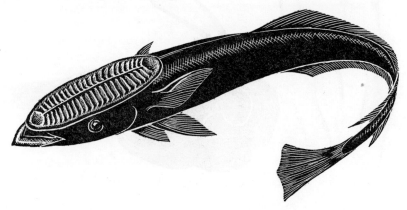

**Figure 9** An echeneid showing the sucker-like dorsal fin.

Some species of echeneid (e.g., *Remora remora* and *Remoropsis pallidus*) are known to augment their food supply by eating the copepod parasites of their hosts, and are frequently found in the mouth and gill cavities where parasitic infestation is often heavy. Other remoras might also obtain part of their nourishment by "grooming" their associate in this way, since the outer row of teeth in the upper jaw form a sharp and narrow blade inclined forwards, ideally aligned for this method of feeding. Such cleaning activities will be dealt with more fully in a subsequent chapter.

### Other implications of transport

We may touch here upon a few cases in which transport plays an important role, though not in the rather specialized sense that we have so far considered. The curious pelagic nudibranch *Phyllirrhoë bucephala* often has attached to it a small medusa (Fig. 10, *bottom*). The latter is now known to be the dispersive

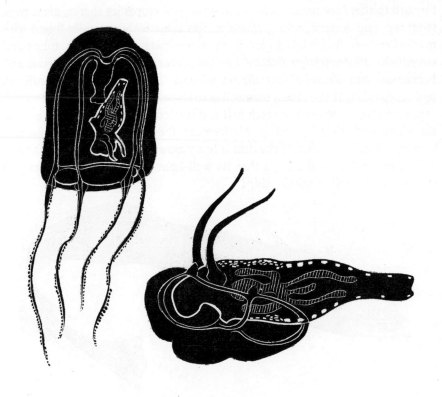

**Figure 10** *Phyllirrhoë bucephala* and *Zanclea costata*. *Top*, the young nudibranch invades the bell of a medusa. *Centre*, eating away portions of its host, it gradually outgrows the latter. *Bottom*, the medusa is reduced to a shrivelled remnant on the now mature nudibranch. (Adapted from a drawing by R. Martin and A. Brinckmann.)

phase of the hydroid *Zanclea costata*, and it was formerly thought to be a direct parasite on the nudibranch. The fact of the matter is that precisely the opposite takes place, the tiny larval *Phyllirrhoë* attaching itself to the medusa's bell and feeding parasitically by sucking at the manubrium and the canal system. As soon as it has outgrown its coelenterate host and is able to swim, it devours both manubrium and tentacles. It would seem that this is a specific association, obligatory for the snail, and that the life cycles of nudibranch and

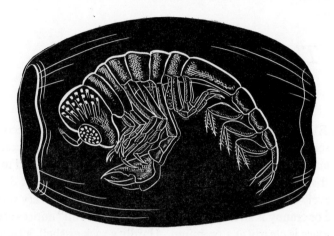

**Figure 11**   The amphipod *Phronima* within the transparent body of its dead tunicate host. (Adapted from various sources.)

hydroid are so timed that the chances of *Phyllirrhoë* establishing contact with its host at the appropriate stages of their development are high (Martin and Brinckmann, 1963).

The so-called barrel-shrimps (amphipods of the family Phronimidae) are found inside the transparent pelagic tunicates known as salps, and in these floating houses the young phronimids are reared. Like the last, this is most probably a parasitic association, *Phronima* slowly eating its salp host, since tunicates containing the amphipod are generally dying or dead (Fig. 11). It is at once apparent that a triple advantage accrues to *Phronima* from this partnership—shelter for itself and its young, an abundant supply of food and wider dispersal for the species.

Dispersal would likewise appear to be the keynote in many other associations involving transport—for example, certain larval stages of some parasitic isopods are frequently found clinging to pelagic copepods. For these, as for other sedentary or parasitic forms, temporary utilization of a "host vehicle" must have considerable survival value.

# 4  *The Hangers-on: Epizoites*

HOWEVER crowded the land-surface of our planet may become in the foreseeable future, one would imagine that the enormous oceanic spaces could never be overpopulated. Surely in the sea there is room and to spare for the myriad species of marine animals. However, even a modest dredge-haul from relatively shallow depths will quickly convince us that every suitable surface, niche or cranny is already occupied—the "house full" notices are out, and available space is clearly at a premium. Hardly a rock or shell, a frond of alga or a discarded beer bottle fails to support a richly varied community.

Of course, much of this colonization is obviously random—animals whose larval stages are not particularly selective can settle happily on a wide range of substrates. Others show strong preferences for one special type of rock or weed, while others again will only complete their development if they can establish contact with some particular animal, on the surface of which they thereafter live. This latter group is the theme of the present chapter.

These epicoles, epibionts or epizoites are often very narrowly specialized forms. As yet we know little about their habits and ways of life, and almost nothing concerning the reasons underlying their choice of rigidly defined substrates. Equally we can only speculate as to the incredible precision with which these favoured sites are discovered by the searching larvae, though certain indications as to how this is achieved are now becoming apparent.

## Unattached epizoites

Although many epizoites are fixed forms, quite a number are unattached and can roam freely over the surface of their host. For these, it can be presumed that the host offers a favourable and precisely defined set of micro-environmental conditions, probably linked with specialized feeding habits on the part of the epizoite. Thus the small, curiously modified annelid *Histriobdella homari* lives among the eggs or in the gill chamber of lobsters, perhaps in a scavenging capacity. The nemertean *Polia involuta* can be found in the egg-

**Figure 12** *Udonella caligorum.* Three small flukes clinging by their posterior suckers to the egg-string of a parasitic copepod (*Lepeophtheirus* sp.) which infests fish. The largest fluke is about 1·5 mm in length. (Adapted from a drawing by V. Dogiel.)

mass of crabs, but apparently feeds only on eggs that are moribund. Small crabs of the genus *Planes* have been observed on turtles, and it has been suggested that they may feed on the reptiles' excrement. *Planes*, however, has been recorded on many other floating objects as well. It is quite possible that many of the highly adapted copepods which dwell on a variety of invertebrates should also come under this heading. A brittle-star, *Nannophiura lagani*, found among the spines of certain sea-urchins, appears to be modified for an epizoic mode of life. Its arm-spines are provided with hook-like processes which aid progression over the surface of its host. This ophiuroid is also remarkable in showing sexual dimorphism, the males being very small.

Although usually described as a hyperparasite, the little monogenetic trematode *Udonella caligorum* could with equal validity be regarded as a narrowly specialized epizoite. This tiny fluke occurs, often in clumps, on caligoid copepods which are themselves parasitic on fish. Generally the egg-strings of the copepod are the site preferred by *Udonella* (Fig. 12). Likewise it seems possible that the whale-louse *Cyamus*, a much modified amphipod, belongs to this category. It is true that the flattened form and powerful claws of *Cyamus* suggest ecto-parasitic status, as does the fact that it is often found around the craters left by whale-barnacles. However, its modifications could equally imply an epizoic existence, and its feeding habits are still obscure. Many whales, especially after visiting Antarctic waters, are covered with a slimy film of diatoms, and it may be that this film (or débris trapped by it) affords some nutriment to these curious amphipods. Certainly whale-lice can be astonishingly successful, over 100,000 individuals being recorded from a single specimen of the grey whale.

### Fixed epizoites

Let us turn now to some fixed forms. The Chonotrichida are ciliate proto-zoans which are attached to the host by one end of the cell while the other end is drawn out either into a feeding funnel or into unequal, paired lips. With a single exception, chonotrichs are epizoic on higher crustaceans. One well-defined group is fairly rigidly confined to the thoracic or abdominal appendages of the host, while another occurs exclusively on the mouth-parts of crabs (Fig. 13). Unlike many other epizoic ciliates, however, chonotrichs are always aligned along the paths of débris currents from the host's mouth. The reason for this seems to be that the feeding-vortex which they produce is small and feeble, and is adequate only if operating in a débris stream (Mohr, 1958).

Suctorian protozoans also figure largely as epizoites. *Hypocoma ascidiarum* can be found around the branchial siphon of sea-squirts, while copepods such as *Notodelphys*, living in the pharynx of ascidians, are often covered by an exuberant growth of *Acineta tuberosa*. Presumably the current set up by the sea-squirt is advantageous to the feeding activities of the suctorians, and this is probably also the reason for the presence of the hydroid *Entocrypta huntsmani* in the branchial cavity of some solitary ascidians. These polyps, however, are known on occasion to ingest ascidicolous copepods (Illg, 1958).

Hydroids seem particularly prone to forming epizoic associations with other animals. We have already noted the occurrence of *Hydractinia* on the shells of hermit crabs, and of *Podocorella minoi* on a scorpaenid fish. In the Indian Ocean, *Nudiclava monacanthi* is found on the skin of a filefish, and in South African waters *Hydrichthys boycei* lives on at least three species of fish. This hydroid, however, has slipped over the boundary into true parasitism, as will be seen later.

**Figure 13**   Two individuals, each about o·1 mm long (excluding the stalk), of the chonotrichous protozoan *Chilodochona quennerstedti*, which occurs on the mouth parts of crabs. (Adapted from various sources.)

Some hydrozoans show a considerable degree of specialization both as to substrate and habit of life. The hydroid stage of the large and beautiful hydromedusan *Neoturris pileata* (Fig. 14) has recently been identified in the Firth of Clyde, where it lives exclusively on the shells of protobranch bivalves belonging to the genus *Nucula* (Edwards, 1965). It is abundant on *N. sulcata*, less so on *N. turgida*, and scarce on *N. tenuis* and *N. nucleus*. These nut-shells are shallow burrowers, most of them making horizontal explorations through the substrate. When the latter consists of fine mud, this appears to provide optimum conditions for the hydroid, which is, curiously enough, adapted to living actually beneath the mud. An allied nut-shell, *Nuculana minuta*, is a much less mobile form and therefore unsuitable as a host for *Neoturris*, though a related hydroid, *Leuckartiara octona*, may occur on its shell. *L. octona* has much longer stems, and is suited to life at the mud surface.

A further example from the Hydrozoa of a remarkably restricted niche is provided by *Ichthyocodium sarcotreti*. This polyp, which completely lacks tentacles, attaches itself only to the parasitic copepod *Sarcotretes scopeli*, which in turn infests the fish *Scopelus glacialis*. Another interesting preference is that shown by certain species of the zoanthid anemone *Epizoanthus*, which regularly occur on what would seem to be a most difficult site to colonize—the siliceous root of the glass-rope sponge *Hyalonema*.

The lovely fan-worms of the family Sabellidae are fairly common on most British coasts, and in certain areas they are associated with a very small but

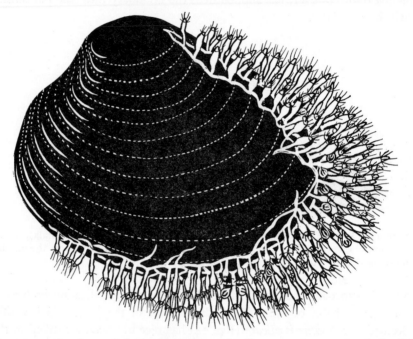

**Figure 14**   A colony of the hydroid *Neoturris pileata* growing on the shell of the bivalve *Nucula sulcata*. (Adapted from a drawing by C. Edwards, by permission of the Council of the Marine Biological Association of the United Kingdom.)

fascinating hydroid, *Proboscidactyla stellata*. The minute zooids of *P. stellata*, barely a millimetre in length, arise from a much-branched stolon ramifying around the mouth of the worm's tube (Fig. 15). They possess only two arm-like tentacles, and bear an absurdly accurate resemblance to tiny human figures—an impression strongly enhanced by their continual bowing and waving movements. The feeding activities of a closely related species have been studied in California (Hand and Hendrickson, 1950). The gastrozooids or food-gathering individuals not only use the ciliary currents set up by the host, but caress the worm's branchial filaments and palps with the highly manoeuvrable tentacles, in this way removing small particles of food.

Worm eggs have also been identified in the gastrozooid's enteron, indicating considerable versatility of feeding habit.

It is a pity that the law governing priority of nomenclature has had to be invoked in the case of *Proboscidactyla*, since Gosse, one of the earlier marine biologists, christened the species *Lar sabellarum*. This beautifully apposite name is derived from the old Roman "Lares" or household gods, and the circlet of tiny zooids, energetically bowing and tossing their arms, could well represent guardian figures at the entrance of their host's dwelling.

**Figure 15** The tiny hydroid *Proboscidactyla stellata* growing around the mouth of a peacock-worm's tube. Each individual is barely 1 mm in length. (Adapted from a drawing by P. H. Gosse.)

An epizoite frequently encountered on the shore is the entoproct *Loxosomella phascolosomata* (Fig. 16). Almost any reasonably sized specimen of the sipunculid *Golfingia vulgaris* will be found to carry several of these entoprocts on its tail end, though they have also been recorded from *Epilepton* and *Mysella*, small bivalves often present in *Golfingia* burrows. What advantage *Loxosomella* gains from occupying this site is conjectural—perhaps the faeces of *Golfingia* are attractive as a food-source, although the anus is remote from the entoprocts' chosen position.

It is also difficult to determine the benefits accruing to certain barnacles which regularly partner sedentary animals. Thus some species of *Scalpellum* occur very often on the stems of hydroids. Others have more exotic preferences, such as *Scalpellum nymphocola* for the legs of a sea-spider living at a depth of nearly 4,000 feet! There is also the case of the acrothoracid barnacle *Trypetesa lampas*. This burrows into the central support or columella of large gastropod shells occupied by hermit crabs, leaving only a small slit by way of communication with the outside world. *Trypetesa* males are dwarfed, and the appendages of the female are rudimentary. The niche occupied, as well as the barnacle's anatomy, suggest extremely specialized feeding habits but, as in so many other instances, we have few clues as to their exact nature.

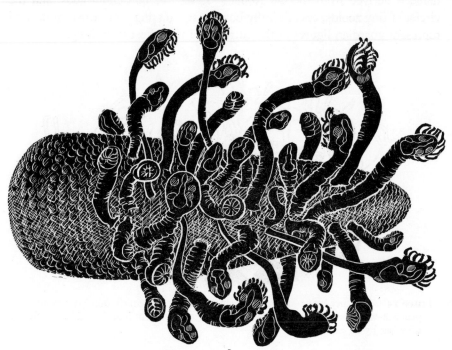

**Figure 16** The tail end of the sipunculid *Golfingia vulgaris* carrying several specimens of the entoproct *Loxosomella phascolosomata*, the largest about 2·5 mm in length.

## Substrate discrimination

Some of the examples given above emphasize the almost unbelievably accurate powers of substrate discrimination which many epizoites must possess. Evidence is now accumulating that recognition *at the molecular level* of appropriate surface texture may be involved (Crisp, 1967). For instance, the attraction which whelk shells hold for the anemone *Calliactis parasitica* has been shown to reside in the periostracal layer—the horny outer covering of the shell. This layer consists of an insoluble protein, apparently detected by the anemone via a highly selective tactile sense. At all events, contact with periostracal tissue triggers a remarkably complex chain of behavioural responses, which result in *Calliactis* climbing firmly on to the shell. The protein concerned is somewhat akin to those known as the arthropodins, found in the cuticle of arthropods, and the presence of which is an important factor in promoting the settlement of barnacle larvae. The latter do not respond to arthropodin in solution, but are evidently "aware" of its presence when adsorbed on a surface—hence the well-known tendency for barnacle cyprids to settle on the shells of older individuals. It is thought that receptors concerned with the recognition of insoluble arthropodin are perhaps present in the cyprid's antennular sucker. If this is so, a sort of Lilliputian lock-and-key mechanism may be operative, a specific molecular pattern eliciting the ultimate response of settlement. Crisp has furthermore suggested that certain specialized epizoic barnacles may have evolved because their present hosts fortuitously possess a surface protein with a molecular structure close to that of arthropodin.

Evolutionary refinement to this level of discrimination may well seem almost too good to be true. We should remember, however, that it is probably no more remarkable than the performance of some male moths which pick up and respond to airborne scents (pheromones) released in minute quantities by virgin females several hundred yards distant—indeed these phenomena may be strictly comparable. The crudity of our own abilities in this respect is such that we can but dimly apprehend the sensory precision of the epizoite's world.

# 5 *The Laundrymen: Cleaning Associations*

FOR many years it has been recognized that certain associations exist in which one animal cleans or "grooms" another, removing and eating its parasites. In this way, the birds known as ox-peckers constantly attend a variety of the larger African mammals, deriving at least a part of their food from the ticks and insects which infest their hosts, and, by characteristic behaviour patterns, warn the latter of impending danger. Only within recent years, however, has the enormous scale of this activity been realized, especially in the sea, where the technique of skin-diving has brought to light many fascinating aspects of this type of behaviour. The term "cleaning symbiosis" has been coined to describe such activities, but we should note that "symbiosis" is here used in its more general sense of "living together". What we are actually witnessing is clearly a very specialized form of mutualism.

## Fish as cleaners

To date over forty species of fish are recognized as cleaners, and many more will probably be described as such in the future (Feder, in Henry, 1966) (Fig. 17). In addition, six species of shrimp are known to practise this habit, as well as a Galapagos crab. This crab removes ticks from the marine iguanas which are only found on these remote islands.

Although cleaning symbiosis certainly occurs in temperate seas, the phenomenon reaches its climax in tropical waters. Here the cleaners tend to be sharply characterized, showing positive modifications of structure, colour and habit, while the clients exhibit behavioural responses often of complex type.

One cleaning species which has been intensively studied in southern

**Figure 17**  A "cleaner" fish, the neon goby (*Elecatinus oceanops*) investigates a large black angelfish (*Pomacanthus paru*). (Adapted from Cleaning Symbiosis by C. Limbaugh, by permission of Scientific American, Inc.)

California is a wrasse (*Oxyjulis californica*), known locally as the señorita
(Limbaugh, 1961). This small brown fish nibbles off the parasitic copepods
and isopods which infest such species as the opaleye (*Girella nigricans*), the
blacksmith (*Chromis punctipinnis*), the topsmelt (*Atherinops affinis*), the black
sea bass (*Stereolepis gigas*), the sunfish (*Mola mola*) and, among elasmo-
branchs, the bat ray (*Holorhinus californicus*). It will also clean away the
white growths resulting from bacterial infection which is a common fish
ailment at certain seasons. The presence of a señorita frequently induces
dense schooling behaviour among its clients, which excitedly mill around the
cleaner. Those actually undergoing the grooming process will often remain
motionless in curious attitudes (sometimes even upside down) while the little
wrasse works over their bodies. *Oxyjulis* appears to enjoy complete immunity
and can safely enter the mouth of large predatory fish in pursuance of its
cleaning activity.

Many species of cleaner fish show convergent adaptation in their pointed
snouts and tweezer-like teeth. Especially in tropical seas, they also tend to
have bright colours and patterns, and are thus extremely conspicuous forms.
This effect is enhanced by special movements and displays which serve to
attract numerous clients to the area. Some practice the cleaning habit only
as juveniles (for example, the grey angel-fish, *Pomacanthus aureus*) and all
seem to have alternative sources of food, though certain species are more
dependent than others on the rewards offered by this type of association.
Another interesting fact which has emerged is that cleaners take up definite
stations marked by salient features on the sea bed—rocky outcrops, coral
heads, wrecks, etc. These stations are well known to the customers which
congregate regularly in their vicinity, queuing up, as it were, to await the
attentions of the cleaner. An astonishing number of fish may be groomed at a
single station—up to 300 have been observed during a six-hour period. We
have already noted that at least some of the remoras are known to clean their
hosts, and it has been suggested that pilot fish associated with manta rays may
do likewise.

That cleaning symbiosis plays an important part in promoting local aggre-
gations of fish is shown by the experiment of removing all the known cleaning
organisms from a single station. Within a fortnight most of the fish which
normally frequented this area had disappeared, and those which remained
rapidly lost condition. Population movements associated with cleaning, however, are by no means restricted to coastal fish—the black sea bass and sunfish
cited above appear to move shorewards specifically for this purpose. It is
obvious that this form of associative behaviour has major implications in the
study of fish ecology.

## The cleaning shrimps

Several species of shrimp offer us another glimpse of cleaning activity in the sea. Of these, the California cleaning shrimp (*Hippolysmata californica*) is the only representative from temperate waters, and seems to be a facultative cleaner. These shrimps roam the sea bed at night in large troops and seek cover in crevices during the daylight hours. They work thoroughly over the surface of any animal encountered, removing epizoites, parasites and decaying tissue. Occasionally they come to grief through incautiously entering the mouth of a moray eel, but in general their activities are tolerated by the host animal.

Much more specialized is the Pederson shrimp (*Periclimenes pedersoni*) from the West Indies (Fig. 18). Unlike *Hippolysmata*, it is a solitary form with somewhat sedentary habits. In our present context it has a double interest, since it invariably lives in company with the sea-anemone *Bartholomea annulata*. This little shrimp is vividly coloured with white stripes and violet spots, and possesses immensely long antennae. These feelers are waved energetically whenever a fish swims by, while the body is swayed to and fro. If a fish requires the attention of *Periclimenes* it will stop a short distance away, and present the appropriate part of its body to the shrimp. The latter thereupon crawls over it, industriously removing parasites from the gills and other affected areas, and cleaning the tissue around injuries. It is even permitted to make small incisions in order to extract subcutaneous parasites, and can enter the host's mouth in perfect safety. As in the case of cleaner fish, the cleaning stations of this shrimp are well known and regularly visited by the local fish population.

## Evolution and the cleaning habit

The widespread occurrence of cleaning symbiosis, so recently recognized, must inevitably have had significant evolutionary consequences. This is well shown by the existence of non-cleaner species of fish which mimic the cleaners in colour and general body form, thus securing for themselves that immunity from predation enjoyed by their models. Nature seldom misses a trick, and has in this case pushed matters a stage further. *Labroides dimidiatus* is a cleaner species which has evolved for display purposes a sort of dance in which the posterior part of the body is moved up and down. This advertises its presence to potential clients who promptly take up invitation postures, thus signifying their readiness to be groomed. Now the size, form and colour of *Labroides*, and even its dancing movements, are simulated by a blenny,

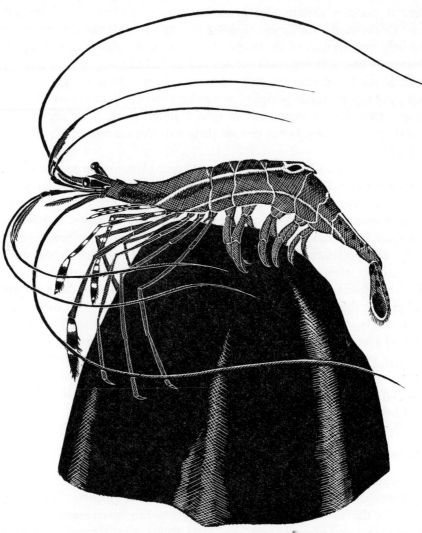

**Figure 18**   The Pederson shrimp (*Periclimenes pedersoni*), one of the more specialized of the "cleaner" shrimps.   (Adapted from various sources.)

*Aspidontus taeniatus*, which thereby elicits an invitation to clean. This blenny is, alas, a wolf in sheep's clothing, being a fin-eater which savagely attacks the deceived clients (Wickler, 1965). We may observe, however, that this subtle evolutionary gambit is not, as yet, perfected. The resemblance between model and mimic is only good enough to deceive younger fish—older and more experienced individuals can apparently tell the difference!

In more general terms, the evolution of the cleaning habit suggests some interesting questions. How, for example, is the predatory urge of a large carnivorous client inhibited? Did certain cleaners start with the initial advantage of unpalatability? Are the bright colours, striking patterns and elaborate rituals of attraction in reality much modified threat displays—a one-time bluff transmuted to a new habit of life? How long has it taken the customers to evolve the quiescent behaviour and curious postures which facilitate the laundrymen? One thing seems sure: the more committed a cleaner has become, and the more dependent for its sustenance on this activity, the greater is its degree of immunity. Once again, we are confronted with the extraordinarily efficient results of long-term selective processes.

# 6 The Messmates: Commensalism

COMMENSALISM in its strict sense denotes a food-sharing association, although in the great majority of cases only one partner acts as the provider. Instances of this are so widespread in the sea, and are of such diverse type, that the chief difficulty lies in the selection of representative examples. It will become apparent, however, that although the acquisition of food due to the host's activities is probably the main feature of these partnerships, additional benefits may be gained by the commensal. Thus such factors as shelter and the provision of a respiratory current may well assume considerable importance.

## The hermit and the worm

Let us start with a classic and much-cited case—that of the polychaete worm *Nereis fucata* and various species of hermit crab. In the Irish Sea, this worm lives in whelk shells inhabited by *Eupagurus bernhardus*, though *Anapagurus laevis* and *Eupagurus prideauxi* have also been reported as hosts. By inducing both worm and hermit to accept an artificial shell of glass, it has been possible to study this association in some detail (see Caullery, 1952). *N. fucata* generally occupies the upper whorls of the shell, maintaining a water current by pulsations of the body. When the hermit is feeding, the worm glides forward and may remove a fragment of food from the crab's mandibles before retreating hastily into the depths of the shell (Fig. 19). Although *Eupagurus* appears to tolerate its large commensal, it would seem in reality to have little choice—food-stealing by this nereid may be reluctantly accepted as a matter of course. Let us not, however, be deceived by the apparent simplicity of this partnership, for there can be no doubt that evolution has acted with considerable success in moulding the nereid's behavioural reflexes—probably even in its pre-commensal stage. For example, the young worm constructs a small tube on the sea-bed, but if the substrate vibrates to the characteristic bumping passage of a hermit's shell, the worm partially emerges to make searching movements, thus contacting its host. The direct nutritional piracy of the adult

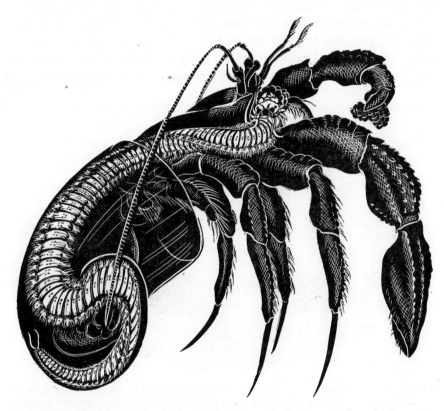

**Figure 19** The worm *Nereis fucata* about to rob its partner hermit *Eupagurus bernhardus*. The pair are living in a glass shell. (Adapted from M. Caullery, after G. Thorson.)

is paralleled by the isopod *Cymothoa praegustator*, which lives in the mouth of its fish host, the menhaden, and steals small particles of food as they pass towards the host's oesophagus.

## Crabs and their anemone partners

Other examples which have received some attention involve hermits and sea-anemones. *Calliactis parasitica* (already mentioned under epizoism) is a pale coloured anemone which frequently adorns the borrowed shells of hermits in many parts of the world (Fig. 20). Sometimes several individuals occur on a single shell. The anemone can, however, live by itself (although association with pagurids is usual), so that commensalism is still to some extent facultative. In some, though not in all areas of the anemone's range, the initiative lies with the crab, which may deliberately plant the actinian on its house. As hermits grow, the existing abode must of course be exchanged for a larger shell, and once this is found the anemones are carefully transplanted. The latter react if their landlord is removed from his shell by spontaneously detaching themselves. Moreover, they tolerate any transplantation assistance

**Figure 20**   The anemone *Calliactis parasitica* with its hermit crab host. This posture, with tentacles trailing behind, is a typical one. (Adapted from various sources.)

which the crab may give without the excessive contraction that might be expected under such circumstances. Undoubtedly life with a pagurid represents the optimum niche for *Calliactis*, since the hermit's movement probably stirs up items of food which can be utilised by the anemone. In return, the strongly developed nematocyst batteries might afford at least a modicum of protection to the crustacean. It is worth remembering, however, that this by now quite complex partnership may well have arisen due to a single circumstance which we have already noted—namely, that *Calliactis* has a marked predilection for living on whelk shells, whether or not they are occupied by a hermit. This is certainly the case in British waters, where the crab host (*Eupagurus bernhardus*) apparently plays no part in establishing the association.

Other pagurid-anemone teams have been described from most of the world's oceans. One which merits particular attention is that of *Eupagurus prideauxi* and the beautifully coloured *Adamsia palliata* (Fig. 21). This is an obligatory association, the two as adults never being found apart. *E. prideauxi* is a small, active hermit, common in British waters, which seems fated to choose a shell that is a little too small to contain it. Once *Adamsia*

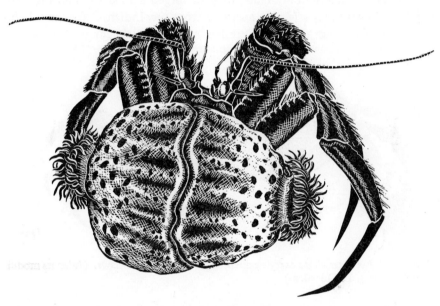

**Figure 21**   The hermit *Eupagurus prideauxi* enveloped by its commensal anemone *Adamsia palliata*. The edges of the expanded pedal disc meet over the crab's back, and part of the circlet of tentacles can be seen on either side. (Adapted from a drawing by T. Stephenson, by permission of the Ray Society.)

has settled, however, the crab's troubles are largely over. The anemone's basal portion steadily expands to form a flexible covering for its partner, thereby allowing for the growth of the hermit, which in its later stages is absolved from the hazards of shell-swopping—always a dangerous process for these crabs with their soft unprotected abdomen. *A. palliata* positions itself so that its mouth opens just ventral and posterior to that of its host—an ideal posture for a food-sharing relationship. Moreover, it has recently been confirmed that the hermit will, on occasion, deliberately place food among the anemone's tentacles (Fox, 1965).

This is obviously an example of highly specialized mutualism, since the adaptations even include certain physiological peculiarities. Indeed, without unduly straining our definitions, the association could almost rank as a true symbiosis. *E. prideauxi* is immune to the toxins which can be extracted from *Adamsia,* though these are fatal when injected into other crabs (Baer, 1951). One might speculate that in the past, this anemone may have attempted a similar commensalism with other hermit species, ultimately finding that only *E. prideauxi* possessed the necessary physiological qualifications in the shape of suitable antibodies.

Hermits are not the only crabs to have forged a food-sharing link with actinians. Two species (*Lybia tessellata* from the Indian Ocean and *Poly-dectes cupulifera* from the Pacific) utilize anemones belonging to the genera

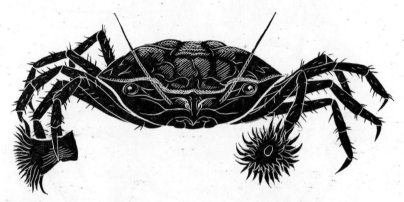

Figure 22   *Lybia tessellata* carrying anemones in its modified claws. (Adapted from a drawing by J. E. Duerden.)

*Bunodeopsis, Sagartia* and *Phellia* in a truly astonishing manner. The claws of the first walking legs are modified so that an anemone can be carried in each, giving the crab the appearance of wearing boxing gloves (Fig. 22). The anemones are employed not only for defence, being thrust forward aggressively if the crab is disturbed, but also as food catchers. So specialized has *Lybia* become that it uses the second pair of legs to push food towards its mouth, the first pair (normally employed for this purpose by crabs) being fully occupied as anemone-holders.

It is interesting to turn briefly to a case in which the anemone rather than the crab initiates the partnership. On the Chilean coast, *Antholoba achates* is regularly found on the carapace of *Hepatus chilensis*. If detached, the anemone responds to contact with the crab's limb, using the pedal disc to move slowly up the leg of its host. When it has reached its customary position on the crab's back, it once more settles down.

## The versatile pea-crabs

The pinnotherids or pea-crabs have been known since very ancient times to associate with other marine animals. More recently they have been recognized as commensals, though in some the balance may tilt in the direction of parasitism. A large number of species have been described and the hosts include bivalves of various sorts, gastropods, sea-slugs, chitons, tube-dwelling polychaetes, echinoderms, burrowing crustaceans and sea-squirts. In a few instances, the crab lives on the external surface of its host. This is the case with *Dissodactylus nitidus*, which is found on the undersurface of the sand-dwelling cake-urchins. Mostly, however, their relations with the partner species are of a more intimate type, frequently involving residence within a cavity of the body. This they are enabled to do because of their small size and

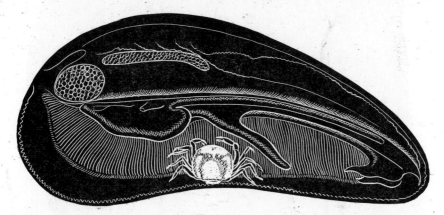

**Figure 23** A female pea-crab, *Pinnotheres pisum*, in an opened mussel, *Mytilus edulis*.

rounded shape, though in *Pinnixa longipes* the body is drawn out laterally so that the crab can fit into the narrow tubes of its worm hosts *Clymenella* and *Pectinaria*.

The technique of cutting small windows in mussel shells (Orton, 1921) has brought to light some interesting features of pinnotherid biology. In *Pinnotheres pisum* (Fig. 23) common in European waters, the food strings of the host are picked from the margins of the gills by the crab's chelipeds. There is some evidence, however, that the very young stages of pea-crabs may simply filter food from the water current produced by the partner, and this almost certainly happens in the adult *P. pinnotheres* when it frequents the atrial chamber of sea-squirts. *P. pisum*, a regular associate of the mussels *Mytilus* and *Modiolus*, is a particularly interesting species, in that recent work has

**Figure 24** A pair of porcelain crabs, *Polyonyx macrocheles*, share the tube of the worm *Chaetopterus variopedatus*. (Adapted from Animal Ecology, by A. S. Pearse, by permission of the McGraw-Hill Book Company.)

revealed a change of host at a certain stage of the life cycle (Christensen, 1958). After several planktonic zoeal instars, the tiny crab, spanning a bare milli-metre in carapace width, invades the bivalve *Spisula solida*. After some time, however, it leaves this host and seeks out a mussel in which to complete its development.

*P. ostreum*, which occurs in the American oyster *Crassostrea virginica*, is known to erode the gills of its partner, thereby causing the latter to lose condition. This pea-crab, therefore, could perhaps be considered a parasite rather than a commensal.

Pinnotherids are often found in the tubes or burrows of other animals. *Scleroplax granulata*, for example, is one of the many commensals which find ideal living conditions in the burrows of the echiuroid *Urechis* and those of the prawns *Callianassa* and *Upogebia* on the North American coast (Fig. 5). *Scleroplax* feeds by straining the incurrent water, though it can also take advantage of larger food particles. On the eastern seaboard of the United States the commodious U-shaped parchment tube of the polychaete *Chaetop-terus variopedatus* shelters another pea-crab, *Pinnixa chaetopterana*, as well as a porcelain crab, *Polyonyx macrocheles* (Fig. 24). The latter belongs to a quite different crustacean group, but shows a certain convergent resemblance to the pinnotherids. Both it and *Pinnixa* feed by filtering the water current main-tained by the worm.

It may easily prove that host-dependence in the pea-crabs is of a double kind. Likely enough not only feeding habits but respiratory needs as well are highly specialized, with the crab's oxygen requirements narrowly geared to the speed and volume of water current produced by a particular host.

## Significance of the host's mode of life

We have already mentioned that the burrows of the mud-dwelling *Urechis caupo* may be shared with fish and pea-crabs. This echiuroid is appropriately named, since the literal translation of its Latin title is "the fat inn-keeper". Apart from the lodgers cited above, another guest is frequently present in the shape of the scale-worm *Hesperonoë adventor* (Fig. 5). *Urechis* secretes a temporary mucus funnel which traps food particles in the inhalant current. When the funnel becomes clogged, it is promptly swallowed and a new one constructed. *Hesperonoë* steals small portions of the food-funnel as well as eating any particles that are too large for its host. It maintains a constant orientation in relation to *Urechis*, and if the latter turns around in its burrow the movement is quickly imitated by the polynoid.

This echiuroid, however, is outdone in the matter of hospitality by the callianassid prawns. On the American coast alone, no fewer than ten associates belonging to widely different groups have been reported, of which about half may actually occur in the same burrow. Another *Hesperonoë* (*H. complanata*) lives with *Callianassa californiensis*, at first lying across the prawn's abdomen, but when larger roaming freely throughout its subterranean home. This polynoid shows marked proprietary tendencies, driving off rivals of the same species, and feeding on scraps swept into the burrow. A type of shrimp, *Betaeus ensenadensis*, may also be found, generally living in pairs. Furthermore, at least two copepod species belonging to the genus *Hemicyclops* must be added to the list of lodgers. An allied burrower, *Upogebia*, likewise seems to attract many commensals.

It is of some interest that *Callianassa affinis*, a close relative of *C. californiensis*, is not known to harbour other animals except for the blind fish *Typhlogobius*. The fact that it prefers to burrow in a much coarser substrate may well contribute to its unpopularity as a host.

A bivalve which utilizes callianassid burrows in a very ingenious way is worthy of mention. This is the small clam *Cryptomya californica*, the siphons of which are very short. On spite of this apparent disability, it can live in sand at depths of half a metre by inserting these little siphons through the wall of the prawn's burrow, and thus filtering the current. *Cryptomya* can also be found, although more sparsely, around the homes of *Upogebia* and *Urechis* (Fig. 5). These latter are not so attractive as hosts, probably because of the more efficient filtration which they themselves carry out.

## Current-producers as hosts

The existence of a water current passing through a host's body is a factor in commensalism, the importance of which can hardly be over-stated. Not only does it frequently satisfy nutritional and respiratory needs, but in many cases it must also act as a sort of chemical beacon to guide the infective stage of commensals in their search for the appropriate partner. This is perhaps one reason why sponges, bivalves and tunicates so often figure as hosts, since the exhalant currents, loaded with metabolic clues as to their origin, can be expelled over quite a distance. Large sponges are especially notable as centres of commensalistic activity, over 16,000 shrimps of the genus *Synalpheus* being recorded from a single loggerhead sponge in the Tortugas. On a smaller scale, a specimen of the sea-squirt *Ascidia mentula*, recently dredged in the Irish Sea and measuring only 7 cm in length, was found to contain the following: two small bivalves (*Musculus* and *Hiatella*); a pea-crab (*Pinnotheres*); an amphipod (*Leucothoë*), and two copepods (*Notopterophorus* and

*Botachus*). Various bivalves likewise afford shelter and food to a large array of specialized copepods, and the big *Cyprina islandica* is host to the nemertean *Malacobdella grossa*. The latter is unusual (for a nemertean) in possessing a posterior ventral sucker, but little is known as to the biology of this association.

One well-known partnership involving a sponge has impinged on the social customs of Japan, off whose coasts the Venus flower basket (*Euplectella*) can be found. Young stages of the stenopid shrimp *Spongicola* enter this sponge to live in pairs in the water passages. After they have grown, escape is no longer possible, and so their elegant prison ultimately becomes a tomb. The beautiful skeletons of *Euplectella* with their contained shrimps are given as wedding presents in Japan to symbolize the permanence of the marriage bond.

## Polychaetes and chemical attractants

A fascinating series of alliances between echinoderms and polychaetes has lately been investigated in some detail (Davenport, 1953 a and b). In particular, scale-worms of the genera *Acholoë*, *Arctonoë*, *Malmgrenia*, *Scalisetosus* and *Harmothoë* all have representatives which characteristically associate with starfish, brittle-stars, sea-urchins and sea-cucumbers. Although detailed information is still scanty, it is clear that some at least of these partnerships are of a food-sharing type. In British waters, *Acholoë astericola* lives in the ambulacral grooves of the sea-star *Astropecten irregularis*, but has on occasion been observed to insert the greater part of its body into the host's stomach and to maintain this position for several minutes. This argues a remarkable immunity to the digestive enzymes of *Astropecten*. In the Mediterranean, the same commensal is found on the large starfish *Luidia*, but although the latter is common around British coasts it seems to be ignored as a host! Ingenious experiments have shown that *Acholoë* responds in some measure to a wide range of starfish on which, however, it is not normally found (Davenport, 1953a). Even isolated pieces of host tissue are attractive to the polynoid, from which we can conclude that a subtle biochemical response is involved.

In the case of *Acholoë*, actual contact with the host appears to be necessary for the attractant substance to become operational. This attractant is rapidly oxidized and is extremely unstable. With the starfish *Evasterias troschelii*, however, the situation is rather different. This asteroid can apparently release sufficient attractant into the surrounding water to summon its commensal polynoid, *Arctonoë fragilis*, from a distance. In the same way, *Hesperonoë adventor* responds to water which has bathed its host *Urechis*, as does the pea-crab *Dissodactylus* to water from the vicinity of its cake-urchin partner, *Mellita*. Under these circumstances, searching behaviour is initiated by the commensal. One very interesting feature of the echinoderm-polynoid associations is the fact that wounded or dying hosts have a repellent effect on the scale-worms. Such a reaction to a moribund partner has of course obvious survival value.

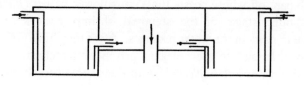

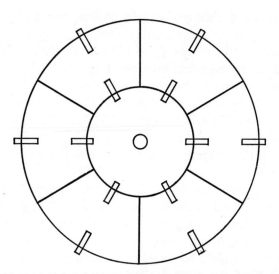

**Figure 25** Two views of a choice chamber designed to test chemical responses in aquatic animals. (Re-drawn from a figure by A. H. Bartel and D. Davenport.)

Many of these investigations on chemical attraction between host and commensal have been carried out with the aid of a "choice" apparatus devised by Bartel and Davenport (1956). Essentially, this consists of a series of radially arranged compartments, each linked via a short glass tube with a raised central chamber which is fitted with a small outlet pipe (Fig. 25). The host is installed in one of the compartments and the commensals in the central chamber. Sea-water is slowly fed at an equal rate into each of the radial compartments. If a chemical attractant is present, it will diffuse into the central chamber and a significant number of the responding commensals will elect to enter the "loaded" compartment, in which they are then trapped. Such relatively simple equipment (and later modifications of it) has greatly increased our information on commensal behaviour.

Another curious problem concerns the occurrence of separate populations of a commensal species on different hosts. The polynoid *Harmothoë lunulata* has been recorded as a partner with thirteen species belonging to four invertebrate phyla, and some research has been carried out on three of these populations (Davenport, 1953b). The results indicate that *Harmothoë* commensal with the burrowing brittle-star *Acrocnida brachiata* respond with equal intensity to the apodous holothurian *Leptosynapta inhaerens*, which can inhabit the same type of substrate. Oddly, however, scale-worms taken from the sea-cucumber do not seem to respond to the brittle-star, although further evidence on this point is desirable. Finally, *H. lunulata* from the terebellid worm *Amphitrite johnstoni* exhibit a sluggish response to *Leptosynapta*, none to *Acrocnida* and occasional or reduced responses to various other polychaetes. Since the scaleworms cohabiting with *Acrocnida* tend to be small, those with *Leptosynapta* somewhat larger, and those from *Amphitrite* larger still, it has been suggested that the three populations simply represent different age-groups. These may, then, respond to different host attractants according to their degree of development. On the other hand, they may be true physiological races, each having a specific sensitivity to one substance or to a "biochemical pattern" emanating from a particular host species. In this case, some degree of attraction to a completely different animal may be due solely to a fortuitous chemical resemblance.

Again, certain polychaetes appear to be divided rather sharply into populations which are either free-living or commensalistic. This happens in the hesionid *Podarke pugettensis*. Worms taken from "free" populations are scarcely attracted at all to animals which in other areas are eagerly colonized (Hickok and Davenport, 1957). Perhaps a similar instance is that of the chlorhaemid worm *Flabelligera affinis* which, over certain parts of its range, associates with the small littoral sea-urchin *Psammechinus miliaris*. Although both are frequently found around the north-eastern coast of Ireland, they do not seem to form a partnership in this region.

Several polynoids also team up with the big tube-dwelling terebellid worms. This is almost certainly a food-sharing association, though it is difficult to visualize at what stage the food is available to the scale-worm. Most terebellids are specialized feeders, nutritive particles being collected by the grooved and ciliated prostomial tentacles which radiate widely from the tube's entrance. Probably the food is pilfered when the particles are collected and consolidated by the lips of the host. One would imagine that the rather spiny scaleworms must be uncomfortable bedfellows for their fat, soft partner within the narrow confines of the latter's dwelling. Once again, a chemical bond between the two has been conclusively demonstrated, and is often of a narrowly specific nature (Davenport, 1953b). Thus the large and beautiful polynoid *Lepidasthenia argus* is strongly attracted to its normal host *Amphitrite edwardsi*, but only weakly to the closely allied *A. johnstoni* and not at all to other terebellids. *Polynoë scolopendrina*, ordinarily associated on the English coast with *Polymnia nebulosa*, appears to have little interest in related terebellids, but responds very strongly to a member of a completely different family, the eunicid *Lysidice ninetta*. Although the latter is known to shelter *Polynoë* in the Channel Islands, it has not been recorded as a host in Plymouth waters (where the association was studied), despite the fact that it is quite common there.

## The finding of the host

At this point, we may reasonably ask how the young stages of commensal animals are guided to a specific host amid the vast complexity of the underwater world. The question is simplified when we remember that evidence has steadily accumulated to show that planktonic larvae do not settle at random, but tend, on the contrary, to exhibit a marked preference for appropriate substrates. If we grant that the host is already present in such areas, the quest is obviously less difficult. It is presumably at this point that the biochemical signposts take over with increasing clarity as the searching larva nears its goal. Even so, ultimate contact and successful metamorphosis may depend on an exquisitely refined appreciation of detail, as we have already seen in certain epizoic species. A further example of this is provided by the larva of the small bivalve *Modiolaria* which has been shown to recognize the molecular structure of tunicin, the substance composing the test of its sea-squirt host (Bourdillon, 1950).

## Specialization in associated copepods

Few classes exhibit so wide a range of partnerships with other animals as do the copepods. In this group a great number of interesting species have recently been described which have food-sharing relationships with a variety of marine invertebrates (Bocquet and Stock, various papers; Gotto, various

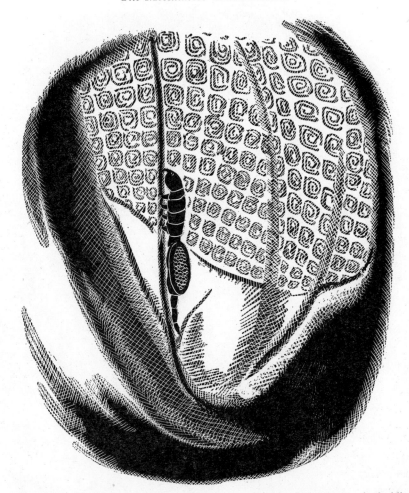

**Figure 26**  An egg-bearing female (about 4 mm long) of the copepod *Ascidicola rosea* clinging to the mucus food-string in the oesophageal region of its ascidian host, *Corella parallelogramma*. The food-string is being slowly drawn by ciliary action into the stomach of the sea-squirt (dark mass at lower right). (Adapted from a photograph by R. V. Gotto.)

papers). Unfortunately, our knowledge of their biology is still very limited, though some details are available for a few species. *Ascidicola rosea* is common around the coasts of Europe, living in various sea-squirts. When it inhabits the remarkably transparent *Corella parallelogramma*, its food gathering and other activities can be easily studied (Gotto, 1957). It clings to the food-string of its host in the oesophageal region and removes small particles from the mucus cord (Fig. 26). As this cord is continually pulled into the ascidian's stomach by ciliary action, the copepod periodically readjusts its position by

C

**Figure 27** Egg-bearing females (about 1·5 mm in length) of the copepod *Sabelliphilus elongatus* on the branchial filaments of the peacock-worm *Sabella pavonina*. (Adapted from a photograph by S. Armstrong.)

climbing up a few millimetres. The eggs of *Ascidicola* are deposited in the host's stomach, from which they travel towards the rectum with the faeces. Hatching takes place as they are expelled from the anus and meet the full force of the exhalant current. The young nauplii are then shot out via the atrial siphon.

*A. rosea* can be regarded as an endocommensal, but another ascidicolous species, *Enteropsis sphinx*, from the colony-forming *Diazona violacea*, more probably achieves true parasitic status. When the time comes for egg-laying, *Enteropsis* migrates from its cramped quarters in the stomach up the long, narrow oesophagus and into the more spacious branchial sac of its host. At all events, only in the latter region can egg-bearing females be found. The interesting feature of this association concerns the whereabouts of *Enteropsis* during the winter months when *Diazona* undergoes a degenerative process, becoming reduced to a mass of buds in a gelatinous matrix. It seems probable that the copepod's immature stages overwinter within these buds.

The problem of successfully evacuating fragile nauplii from the interior of a host specifically designed to trap micro-organisms must be a very real one for the ascidian-dwelling copepods. *Ascidicola*, it is true, utilizes the rectal route, but it is probable that the young of many species leave via the inhalant siphon. Since escape against the steady force of the water current appears unlikely, it is thought that the female copepod may take advantage of the brief periodic reversals of current which punctuate the normal feeding activities of sea-squirts. It may even be that the movements of the female during parturition can themselves induce such alterations of flow, since gentle stimulation within the rim of the inhalant siphon produces just this effect.

The exact site occupied by these copepods is of importance in determining their precise relations *vis-à-vis* the host animal. *Lichomolgus protulae* occurs on the frontal surface of the filaments composing the feeding fan of serpulid worms and is thus in a position to gather nutriment from the food-strings of its partner. Its status, therefore, is presumably that of a commensal. However, a related lichomolgid, *Sabelliphilus elongatus* from the peacock worm, lives on the abfrontal surface of the filaments (Fig. 27) and has been shown to be an ectoparasite, since it erodes and eats the substance of its host's fan (Gotto, 1960).

Many such copepods show bizarre alterations of the mouthparts, suggesting extremely specialized feeding habits. The mandibles of *Eunicicola insolens* (Fig. 28), a species living on eunicid worms, include small fringed scoops, which probably push minute food fragments towards the mouth. *Eunicicola* also possesses one large ventral sucker and four tiny antennal suckers. The former is useful for adhesion on the broad, smooth back of the polychaete,

**Figure 28**   Modification in associated copepods. *Top left*, the worm-like female of *Mycophilus roseus* (1 mm long) from botryllid sea-squirts. *Centre left*, female *Eunicicola insolens* (0·8 mm) from eunicid worms. *Bottom*, the specialized mandible of *Eunicicola* (0·1 mm). *Right*, *Rhodinicola gibbosa* (4·3 mm) from maldanid worms. (*Rhodinicola* adapted from a drawing by J. Bresciani; others from R. V. Gotto.)

while the latter are employed when the copepod moves on to the feathery gills.

Another curious dweller on annelids is *Paranthessius myxicolae*. This copepod is associated with the sabellid *Myxicola infundibulum*, which builds a thick gelatinous tube in the sand. The tube is laid down as a series of concentric rings, and the copepods are usually found imprisoned in the narrow space between two successive layers. It is almost certain that they feed exclusively on the mucoproteins of the tube.

A noteworthy aspect of the biology of associated copepods lies in the varying number of eggs produced from species to species. Some forms carry only two eggs at a time while others may lay several hundred. It is probable that egg-production is closely linked with the copepod's ecology (Gotto, 1962). Thus species which are faced with special difficulties or hazards in the search for their appropriate host are likely to be characterized by a high egg-number.

## Some little investigated cases

Certain regularly-occurring partnerships, although well known, remain somewhat ambiguous. *Echinocardium cordatum*, a burrowing heart-urchin frequently encountered around British shores, is often accompanied by the little bivalve *Montacuta ferruginosa*. The mollusc is found in the "sanitary

**Figure 29** The gastropod *Sipho curtus* carrying on its shell the anemone *Allantactis parasitica*, beneath the pedal disc of which lives the nemertine worm *Nemertopsis actinophila* (seen projecting posteriorly). (Adapted from D. Davenport, after G. Thorson.)

burrow" near the anus of its host, where it presumably benefits from the stream of expelled débris. Again, several annelids have been reported as sheltering in the lateral folds of enteropneusts or acorn-worms belonging to the genus *Balanoglossus*. We may suppose that they derive some advantage from the current of water issuing from the host's gill-slits, but the feeding habits of these polychaetes remain obscure.

Before leaving invertebrate associations, we should mention an extraordinary partnership recently described from Greenland (Thorson, in Davenport, 1955). This is a sort of "triple-decker" arrangement, involving three species rather than the usual two (Fig. 29). The participants are the gastropod *Sipho curtus*, the anemone *Allantactis parasitica* and the nemertean *Nemertopsis actinophila*. The anemone lives on the shell of the sea-snail, while the worm finds shelter beneath the actinian's pedal disc. *Nemertopsis* is permitted to enter the anemone's gastral cavity, from which it removes food, although other worms are liable to be seized and digested. The harmonious integration of three such different animals is certainly a tribute to the power of adaptive specialization.

### Pilot fish and sharks

Finally let us turn to an exclusively vertebrate team—that of the pilot fish *Naucrates ductor* with various species of large sharks and manta rays (Fig. 30). As is well known, these little fish are true commensals, accompanying their

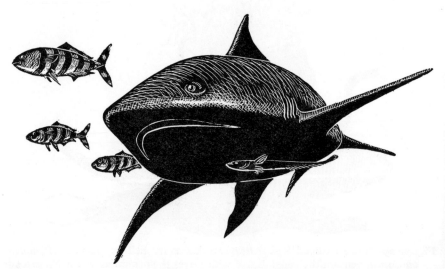

**Figure 30**  A shark accompanied by pilot fish (*Naucrates ductor*). An echeneid adheres to its host's belly. (Adapted from various sources.)

formidable protector wherever he swims and sharing in the food which he obtains. Plenty of scraps are usually available since sharks are messy feeders. Because pilots generally maintain a more or less fixed position near the head of their host, it has sometimes been assumed that they guide the shark towards potential food. It is much more likely that *Naucrates* is simply taking advantage of the small pressure wave set up by the big fish—and in any case, sharks are well equipped to seek their own prey unaided. Their toleration of the pilot fish, however, remains unexplained. It appears that *Naucrates* has a tendency to adopt any large floating object, as shown by the school of pilot fish which accompanied the *Kon-Tiki* raft after their "own" shark had been killed. If a hooked shark is being hauled out of the water, its attendant fish will fret anxiously around the ascending body before swimming off in search of a new protector.

In this chapter we have done no more than touch lightly on the subject of true commensalism. Many more examples are known, and there can be no doubt that a large number of similar partnerships still await description and study.

# 7 *The Intimates: Physiological Symbiosis*

WE have already defined symbiosis as an association involving some loss of physiological independence by both participants, and have cited the protozoan fauna of termites and ruminant mammals.

These instances were given to illustrate the very intimate relationship which symbionts can achieve. Comparable and equally well-studied examples from the marine environment are not so easy to find, although they almost certainly exist. A great deal of interest, however, has centred on the association between certain minute unicellular forms and a variety of marine invertebrates—coelenterates, sponges, worms, molluscs and others. It is somewhat difficult to place these tiny organisms systematically, since many belong to that curious hinterland that lies somewhere between the plant and animal kingdoms. Thus some authors regard them as single-celled algae, others as flagellate protozoans possessing chlorophyll or xanthophyll. It would seem that the consensus of modern opinion is that they are for the most part unicellular algae. Green forms occur mainly in fresh water and are known as zoochlorellae, while the yellow or brown ones (zooxanthellae) are marine. Although some can apparently exist in a free state, they are characteristically found in the tissues of an invertebrate host and are often, in fact, highly specific.

## *Convoluta* and its symbionts

Our first example must be dealt with in some detail since it has figured largely in the annals of symbiosis. *Convoluta roscoffensis* is a small flatworm belonging to the acoelous division of the class Turbellaria. It inhabits the sandy beaches of Brittany and, although only a few millimetres in length, may occur in such numbers as to give the sand a greenish tint. When the tide comes in, these tiny worms burrow down between the sand grains. The green tint is due, not to the colour of the worms themselves, but to the fact that their tissues are loaded with minute green algae. The young flat-worms have a very simple gut by means of which they can ingest the sand-dwelling algae, which promptly invade the rest of the host's tissues. As the worm matures, however, the gut becomes occluded, so that solid food can no longer be taken. In this predicament *Convoluta* turns on its "guests", at first digesting only the surplus. After the flatworm has laid its eggs, the reproductive potential of the algae is outstripped by the rapacity of their host, which now inexorably consumes the entire population. Since it can neither reinfect itself nor feed in an orthodox manner, *Convoluta* thereupon dies. Its end is probably hastened by the fact that it lacks an excretory system, so that the metabolic products which were formerly utilized by the algal associates now accumulate in lethal concentrations. A related species, *Convoluta convoluta*, retains its gut throughout life and can not only feed in the usual way but can also reinfect itself. However, it does not thereby achieve independence, since in the absence of its symbionts, it cannot develop to maturity. Some at least of *C. convoluta*'s symbionts are now reported to be diatoms (Ax and Apelt, 1965).

Although this association is generally classified as a symbiosis, such a description is not strictly accurate. It is certainly true that the alga is physiologically essential to *Convoluta*, but the reverse does not hold since the algae can and do exist in a free-living state, although they are chemically attracted to the worm's egg-cases. At most, the tissues of the flatworm provide them with a supremely favourable environment—at the price of ultimate destruction! As one zoologist has pointed out, the association is really a kind of aberrant mutualism.

### Do reef-building corals require zooxanthellae?

For many years there has been speculation concerning the role of zooxanthellae in the biology of reef-building corals, in the tissues of which they occur very plentifully. Some workers thought that they were utilized as a source of food since, if corals were starved, the zooxanthellae appeared in a partially broken down state in the absorptive region of the host's mesenteries (Boschma, 1926). Further work, however, showed that this was the route by which they were expelled and was not part of the coral's digestive process (Yonge and Nicholls, 1931). Furthermore, there seemed to be virtually no evidence that zooxanthellae could be used as food, since corals are strictly carnivorous, feeding during the hours of darkness on the animal plankton. It was therefore felt that if this association was symbiotic, it could not be concerned with the host's nutrition. However, the photosynthetic activities of the zooxanthellae would undoubtedly supply additional oxygen—at least during the daytime—while their utilization of the host's excretory products (carbon dioxide, phosphorus and nitrogen) would result in the rapid and efficient removal of these wastes. Though the additional oxygen might have little significance in well aerated waters, the prompt elimination of metabolic by-products should certainly promote healthy and vigorous growth. It seemed likely, then, that while zooxanthellae were not vital to corals living under good conditions, their presence in adverse ecological circumstances might well tip the balance in favour of maintaining a colony in a viable state.

It would be fair to say that this was the prevailing view until very recently, but the picture has now been modified as a result of some Japanese researches with the aid of the electron microscope (Kawaguti, 1964a). From this work it appears that, in some corals at least, zooxanthellae are inter- and not intra-cellular associates, as was formerly believed. It has further been demonstrated that layers of a scale-like substance cover the surface of a zooxanthella, which are probably derived from the outer membrane of the latter. This substance decomposes, particularly if the coral is kept in darkness, and is absorbed by the surrounding cells of the host. Preliminary investigations suggest that the substance concerned is of a carbohydrate nature (Kawaguti, 1964b). Future research may well confirm that coral tissues receive nutriment from zooxanthellae in this way, which does not, of course, involve sacrificing the symbionts. It should also be mentioned that some of the soft corals have abandoned the capture of living prey, and so presumably rely entirely on the nutritive substances supplied by their contained zooxanthellae.

There is, however, one further aspect of the coral story that must be mentioned. Goreau (1961) has now established experimentally that the algae contained in reef corals fulfil a most important function other than the possible trophic assistance which they may render to their hosts. By their photosynthetic activities they greatly increase and facilitate the deposition of calcium—an element obviously required in enormous quantity by reef-builders with a limestone skeleton. Goreau indeed believes that this association has been critical in the evolution and maintenance of large reefs.

## The significance of zooxanthellae for the tridacnid clams

If the exact nature of this partnership between corals and unicellular algae still poses certain questions, the case of the large clams comprising the family Tridacnidae is less ambiguous. These bivalve molluscs are characteristic inhabitants of reefs in the warmest parts of the Indo-Pacific (Fig. 31). They

**Figure 31** *Top*, a giant clam (*Tridacna derasa*), its shell valves slightly open to show the extensive mantle tissue around the siphonal area. *Bottom*, a small section of the mantle tissue showing zooxanthellae (white dots) clustered around one of the lens-like structures. (Mantle tissue adapted from a drawing by C. M. Yonge.)

are beautiful and conspicuous objects when viewed through the clear, shallow water, since the exposed soft tissues resemble gorgeously coloured velvet. Peacock-blue and moss-green are common shades, often in combination with other colours. The largest member of the family is the giant clam *Tridacna derasa*, which may attain the astonishing length of six feet, and possesses a pair of massive shells with smoothly scalloped borders. These enormous molluscs contain many thousands of tiny zooxanthellae, especially in the soft tissue which is fully exposed on the upper surface of the clam whenever the latter is submerged. This tissue represents an immense extension of the animal's siphons and of the area immediately surrounding them. In most bivalves, the siphons (through which water enters and leaves the body) are not nearly such conspicuous features and are moreover restricted to a relatively small area at the posterior end. In effect, the tridacnids have undergone a morphological rotation so that this brilliant, hypertrophied tissue is directed upwards. The zooxanthellae are particularly numerous in the blood sinuses which traverse it, where they occur in phagocytic cells within which they can divide.

These marvellously coloured tissues can be regarded as pastures in which the contained symbionts are not only protected but are surrounded by their host's waste products—metabolites which can be readily utilized for the build up of proteins. At this point, however, resemblance to the coral symbiosis ceases, since there is no doubt that the clams digest their tiny partners. This digestion does not take place in the gut, but in the blood system, and presumably accounts for the very large size of the clam's renal organs, which must be adapted to cope with the considerable quantities of waste material resulting from this method of nutrition. The symbiotic algae thus constitute an important source of food, although tridacnids, like many bivalves, also feed on plant plankton which they filter from the surrounding water.

Perhaps the most interesting aspect of this association concerns the evolutionary path which these big clams have taken. The reorientation and expansion of the siphonal area ensures maximum exposure to the hot sunlight which the zooxanthellae require. At the same time, the deep pigmentation of this tissue safeguards the underlying cells of the host from the intense solar radiation. The tridacnids, however, have not been content with merely gross adaptations of this sort. On the upper surface of the siphonal tissue they possess many small lens-like structures (Fig. 31). Originally, these were very probably visual organs, but now serve to focus light into the deeper layers where the zooxanthellae are densely clustered. Moreover, the whole ecology of the Tridacnidae—dwellers exclusively in shallow and very warm water—contributes notably to the efficiency of the symbiosis. It may even be that the gigantic size of *T. derasa* has been made possible because of the additional nutriment represented by its symbionts—an evolutionary bonus from a long-standing and successful partnership.

## The widespread occurrence of possible physiological symbioses

Many other cases are recorded of zooxanthellae living in the tissues of marine animals. In some, the symbionts are acquired, as it were, second-hand, a good example being provided by certain nudibranchs such as *Aeolidiella glauca* which prey on already infected sea-anemones. What role, if any, the uni-cellular forms play in their new host remains uncertain. Again, some modifications of habit on the part of a host animal strongly suggest that a functional symbiosis is operative, as in the case of the tropical jellyfish *Cassiopeia*. Unlike other scyphomedusans, this sluggish, shallow-water form lies on the bottom with its oral surface directed upwards, thus exposing the numerous zooxanthellae living in its sub-umbrellar tissue to the full blaze of sunlight.

As our knowledge increases, it is probable that a fair number of marine associations will be found to involve some degree of physiological inter-dependence. Until more information is available, however, it seems best to restrict our examples to the relatively well investigated cases cited above.

# 8 *The Paths to Parasitism*

A BRIEF assessment of marine parasitism can now be attempted, and of the ways in which it may have arisen from the various types of association considered in the previous chapters. By a parasite, we mean an animal which lives with a host during at least a part of its life-cycle, obtaining successive meals from its body but not destroying it as a predator would its prey. The host is frequently harmed in some way by the presence of the parasite, though such injury is not necessarily permanent, a state of balance often being achieved in an association of this kind.

## Types of parasitism

Several types of parasitism are known. In continuous parasitism, the entire life-cycle is passed on or in a host animal. Temporary parasitism may be exemplified by the small crustacean *Argulus* (the carp-louse) which infests fish, but can swim freely from host to host between meals. In protelean parasitism, only the young stages are dependent on association with some other animal. An excellent example is that of the monstrillid copepods. As an adult, *Monstrilla* is free-living, though it is aberrant in lacking an alimentary canal. The nauplii bore their way into certain species of polychaetes or prosobranch molluscs, and, once installed, develop two long horns through which nourishment can be absorbed. On completing their development, the adults break out through the skin of the host. Quite recently, the nauplii of a related genus, *Thespesiosyllis*, were discovered in the stomach of a brittle-star, but the details of this life-history are still unknown (Bresciani and Lützen, 1962). Another instance of protelean parasitism is that of the isopod *Gnathia*. Like the monstrillids, *Gnathia* lacks a gut, but is a free-living form. Its larval stage, however, attaches itself to fish and gorges on the latter's blood. Just the opposite occurs with the cymothoid isopods—these swim freely as larvae, but when adult attack fish and cephalopods, on which they become ectoparasites.

## Isopods and cirripedes as parasites

It will be worthwhile to review briefly two groups in which tendencies towards a parasitic mode of life become increasingly apparent. The isopods are, in the main, a free-living order of crustaceans. We have seen, however, that some have developed mouth-parts capable of piercing the skin of fishes. This trend towards host-dependence reaches a climax in the sub-order Epicaridea all of whose members are greatly modified parasites of other Crustacea. In the bopyrids, for example, the first larval stage enters the gill chamber of a decapod. If this host is not already infected, the larva develops into a large bloated female with a distended brood-pouch, which produces a distinct swelling on one side of the host's carapace. Any larvae which subsequently reach the same host will become rather small males. In the same sub-order the entoniscids are even more modified as internal parasites of crabs. Here the crab reacts by forming a sort of membranous sheath around the almost shapeless body of the female entoniscid. The male is small and lives in the brood-pouch of the female.

One can trace a similar series in the cirripedes. Some barnacles, as mentioned earlier, are epizoic on whales. Others, such as *Anelasma squalica*, project from the skin of sharks, usually just anterior to a dorsal fin and often in pairs, suggesting that settlement by the first individual promotes the chance of subsequent infestation (Fig. 32). *Anelasma* has poorly developed feeding appendages and a somewhat reduced alimentary canal, but is possibly able to feed in the usual cirripede manner. The embedded portion of the body, however, has a number of rooting processes, and around these the host tissue tends to be liquefied. There is thus some evidence that a certain amount of nutriment is obtained in a parasitic fashion. *Rhizolepas annelidicola* goes a stage further. This barnacle is attached by branching roots to the scale-worm *Laetmatonice* and, although retaining quite well developed appendages, lacks both mouth and anus (Fig. 32). Its food must therefore be derived solely from the worm host.

In these latter cases, the root system appears to originate from the stalk of the barnacle and is relatively simple. Much more complex are the absorptive roots of the rhizocephalan barnacles, which represent the end point in the evolution of parasitic cirripedes. These infest various other crustaceans, and

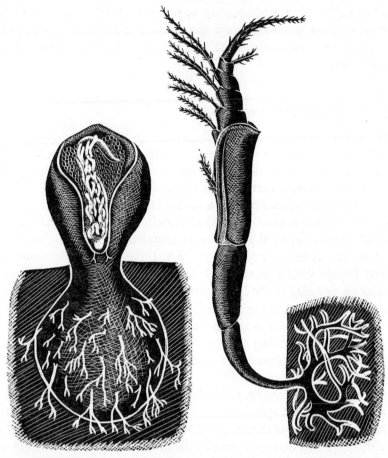

**Figure 32**   The highly modified cirripedes *Anelasma squalica* (*left*) and *Rhizolepas annelidicola* embedded in the tissues of their hosts. (Adapted from J. G. Baer, after drawings by J. Johnstone and W. Frost, and J. H. Day.)

include the well-known *Sacculina carcini*, which appears as a diamond-shaped sac on the abdomen of shore-crabs.  Here the rooting system ramifies throughout the host's body, beginning as a tiny mass of undifferentiated cells derived from the infective larval stage.  Rhizocephalans have lost all resemblance to their free-living barnacle kin, save that their nauplius stages are unmistakably of cirripede type.

### Rarity of parasitism in certain groups

In both isopods and barnacles, therefore, parasitism is sufficiently common for its step-by-step development to be traced with ease.  In some other groups, however, it is surprisingly rare. There are, for example, very few genuinely parasitic polychaetes, though an interesting exception is *Ichthyotomus sanguinarius* which affixes itself to the fins of certain eels and sucks their blood. Although placed in a family of its own, this worm shows resemblance both to the syllids and eunicids (the great majority of which are free-living) in possessing a pair of complex piercing stylets not dissimilar to the jaw apparatus found in these two families. A high degree of parasitic adaptation is apparent in the possession of a sucker-like head area and of anti-coagulant glands.

**Figure 33** The gastropod *Turbonilla elegantissima* feeding on the tentacles of a cirratulid worm. (Adapted from a drawing by V. Fretter.)

The Mollusca are another group which show very little predilection for this mode of life. Out of approximately 100,000 known species only about 100 (0·1 per cent) could be described as parasitic. It is true that certain extremely specialized molluscs—recognizable as such largely by their larval stages—infest echinoderms, but some of these are so grotesquely modified that it is difficult to visualize their starting-point. Of more significance in our present context is the gastropod family Pyramidellidae. *Turbonilla elegantissima* is to all intents and purposes an ordinary little sea-snail with a slender gracefully tapered shell (Fig. 33). It feeds however as a parasite on the long extensile tentacles of the sedentary polychaetes *Amphitrite* and *Audouinia*, as these grope through the silt in search of food particles (Fretter, 1951). The worms' tentacles are held by a small oral sucker and pierced by a stylet, before the food is sucked up the snail's proboscis by means of a buccal pump. Other members of the family have rather similar habits, parasitizing a variety of marine invertebrates.

## The predator–prey relationship as a prelude to parasitic habits

*Turbonilla*'s method of obtaining a living is not altogether unlike that of the dog-whelk *Nucella*, which plunges its proboscis into the soft tissue of barnacles and cleans them out, leaving only the empty shell. The main difference clearly lies in the fact that *Turbonilla* does not kill its victim. Nevertheless, we can see at once how thin is the line dividing a predator–prey relationship from that of parasite and host. Here obviously is a major route towards parasitism as a way of life. We have only to visualize the evolution of an increasingly specialized form of predation, in which the animal providing the meal is permitted to live. This must have happened many times in the past, especially in cases in which the potential parasite is an appreciably smaller animal than its potential host.

## Other evolutionary pathways

What of other associations, and their relevance to the evolution of parasitic habits? We have seen that epizoism may include partnerships ranging from the very casual to the obviously adapted. As mentioned earlier, certain hydroids are characteristically associated with various species of fish. In at least one instance this relationship has become truly parasitic. *Hydrichthys boycei* has been recorded on three South African fishes, occurring on the skin in almost any part of the body (Fig. 34). A colony consists of a basal plate-like portion, from which arise the hydranths and gonostyles. This basal plate grows at its edges, and also sends processes into the body of the fish, which penetrate between the host cells and actually absorb them. The hydranths or feeding zooids are remarkable in lacking tentacles, and possess highly distensible mouths. These mouths can be bent down and applied to the damaged tissue around the edges of the hydrorhizal plate. After a time—probably following the release of medusae—the colony either dies or drops off its host, leaving, however, a noticeable scar.

From the evidence of other fish-hydroid associations, it seems reasonable to assume that *H. boycei* began its career as a simple epizoite. It may be that the superficial erosion of host tissue by the basal plate then provided an alternative source of food and thus triggered a chain of selective modifications which have resulted in the highly specialized and aberrant hydroid which we see today. In a similar manner, the cirripede *Anelasma* almost certainly lived epizoically on the skin of sharks before capitalizing on its rooting processes and opting for the comparative ease of parasitic existence.

**Figure 34** A colony of the parasitic hydroid *Hydrichthys boycei*, with feeding hydranths and reproductive individuals in various stages of development, on the skin of its fish host. *Near top right*, the mouth of a hydranth can be seen applied to the eroded tissue at the edge of the basal plate. (Adapted from a drawing by E. W. Gudger.)

Probably true commensalism, in the sense of a shared food-source, rarely leads to parasitism. It is difficult to imagine that *Nereis fucata*, for example, would ever have been impelled by evolutionary pressures to turn on its hermit host—or not, at least, while the latter obligingly continued to obtain an adequate supply of suitable food. Again, the little crabs from *Chaetopterus* tubes are interested only in the provision of a food-laden current rather than in the gastronomic possibilities of the current-maker. At the same time, we should remember that the primarily commensalistic pea-crabs do show evidence of incipient parasitism in some of their molluscan hosts. It is likely, too, that an originally commensal existence has become something more sinister in a number of associated copepods. Many, like *Notodelphys*, are content to roam the pharynx of their sea-squirt host, utilizing an insignificant fraction of the collected food. Others, however, such as *Haplostoma eruca*, have moved on into the stomach and intestine, and exhibit the tell-tale morphology of parasitic forms. Still others, probably via the original pharyngeal route, have invaded the blood vessels and show every sign of extreme modification—reduction of limbs, vermiform shape, etc. Examples include the worm-like *Scolecodes huntsmani*, up to 14 mm long, from the endostylar vessel of *Pyura haustor*, and *Mycophilus roseus* from the canal system of botryllids (Fig. 28). Copepods which seem to have taken a short cut and become intensely specialized parasites while still in the pharynx are *Gonophysema gullmarensis* from *Ascidiella*, and the cyst-forming *Kystodelphys drachi* from *Microcosmus*.

Purely shelter associations in which mere proximity to the protector is the dominant feature may not constitute an important pathway to parasitism, but the same can hardly be said of inquilinism. An animal which habitually shelters within the body of another achieves an intimate acquaintance with host tissue which, on occasion, may well prove tempting. Provided that its method of obtaining food is not too specialized to start with, a successful change of feeding habit carries with it the bonus of more permanent protection within the host, the latter thus ceasing to be a refuge only and becoming the source of nutriment as well. The Mediterranean pearl-fish, *Carapus acus*, nibbling tentatively at the gonads of its holothurian partner, may be taking this route to parasitic status. It is also quite possible that certain copepods, particularly those belonging to the family *Clausiidae*, started as mere endoeketes in the tubes of their polychaete hosts. Some clausiids have stuck to this mode of life and retain many of the features characteristic of their free-living kin. Others, however, such as *Rhodinicola gibbosa* (Fig. 28), are firmly attached to the body surface of maldanid worms, on which they are almost certainly true parasites. Finally, at least one genus (*Entobius*) lives in the gut of terebellids.

Under certain circumstances, of course, animals which were originally ectoparasites may ultimately become endoparasitic. The lung-mites (*Halarachne*) of seals, for instance, have undoubtedly moved to their present situation from an external site via the nasal passages of their pinnipede hosts.

It is clear, then, that there are more ways than one of becoming a parasite. That so many paths exist is not perhaps surprising when we consider the almost infinite variety of marine organisms, the multiplicity of niches, the permanence of the great oceans and the enormous span of time for evolutionary experiment. In this book we have been able to survey only a few of these experiments in associative existence out of the many already known and still unknown. It is certain that future work will reveal fresh patterns of this sort in the marvellous fabric of marine life.

# References

Good general accounts of various marine associations will be found in the following works:

*Parasitism and Symbiosis*, by M. Caullery. Sidgwick and Jackson, Ltd., London. 1952.
*The Biology of Marine Animals*, by J. A. C. Nicol. Pitman. 1960.
*Symbiosis*, edited by S. M. Henry. Volume 1—Associations of Microorganisms, Plants, and Marine Organisms. Academic Press. 1966.

The references given in the text are listed below.

Arnold, D. C. (1957). Further studies on the behaviour of the fish *Carapus acus*. *Pubbl. staz. zool. Napoli*, **23**, 91.

Ax, P., and Apelt, G. (1965). Die "Zooxanthellen" von *Convoluta convoluta* (Turbellaria Acoela) entstehen aus Diatomeen Erster Nachweis einen Endosymbiose zwischen Tieren und Kieselalgen. *Die Naturwissenschaften*, **15**, 444.

Baer, J. G. (1952). *Ecology of Animal Parasites*. Urbana Univ. Illinois Press.

Bartel, A. H., and Davenport, D. (1956). A technique for the investigation of chemical responses in aquatic animals. *Brit. J. Anim. Behav.*, **4**, 117.

Boschma, H. (1926). On the food of reef-corals. *Proc. Acad. Sci. Amsterdam*, **29**, 993.

Bourdillon, A. (1950). Note sur le commensalisme de *Modiolaria* et des ascidies. *Vie Milieu*, **1115**, 198.

Bresciani, J., and Lützen, J. (1962). Parasitic copepods from the west coast of Sweden including some new or little known species. *Vidensk. Medd. fra Dansk naturh. Foren.*, **124**, 367.

Caullery, M. (1952). *Parasitism and Symbiosis*. Sidgwick and Jackson, London.

Christensen, A. M. (1958). On the life history and biology of *Pinnotheres pisum*. *Proc. Internat. Congr. Zool. London*, **15**, 267.

Crisp, D. J. (1967). Barnacles. *Science Journal*, **3** (9), 69.

Dahl, E. (1959). The amphipod *Hyperia galba*, an ectoparasite of the jelly-fish *Cyanea capillata*. *Nature*, **183** (4577): 1749.

Davenport, D. (1953a). Studies in the physiology of commensalism. III. The polynoid genera *Acholoë*, *Gattyana* and *Lepidasthenia*. *J. mar. biol. Ass.* U.K., **32**, 161.

Davenport, D. (1953b). Studies in the physiology of commensalism. IV. The polynoid genera *Polynoë*, *Lepidasthenia*, and *Harmothoë*. *Ibid.*, **32**, 273.

—— (1955). Specificity and behaviour in symbioses. *Quart. Rev. Biol.*, **30**, 29.

—— and Norris, K. S. (1958). Observations on the symbiosis of the sea anemone *Stoichactis* and the pomacentrid fish *Amphiprion percula*. *Biol. Bull.*, **115**, 397.

De Bary, A. (1879). *Die Erscheinung der Symbiose*. Cassel, Strasbourg.

Dinamani. P. (1964). Variation in form, orientation and mode of attachment of the cirriped, *Octolasmis stella* (Ann.), symbiotic on the gills of lobster. *Journ. Animal Ecology*, **33**, 357.

Edwards, C. (1965). The hydroid and the medusa *Neoturris pileata*. *J. mar. biol. Ass. U.K.*, **45**, 443.

Feder, H. M. (1966). Cleaning Symbiosis in the Marine Environment. Chapter in *Symbiosis* (Vol. I), edited by S. M. Henry (see above in general list).

Fox, H. M. (1965). Confirmation of old observations on the behaviour of a hermit crab and its commensal sea anemone. *Ann. Mag. nat. Hist.* (13), **8**, 173.

Fretter, V. (1951). *Turbonilla elegantissima* (Montagu), a parasitic opisthobranch. *J. mar. biol. Ass. U.K.*, **30**, 37.

Goreau, T. (1961). Problems of growth and calcium deposition in reef corals. *Endeavour*, **20** (77), 32.

Gotto, R. V. (1957). The biology of a commensal copepod, *Ascidicola rosea* Thorell, in the ascidian *Corella parallelogramma* (Müller). *J. mar. biol. Ass. U.K.*, **36**, 281.

—— (1960). Observations on the orientation and feeding of the copepod *Sabelliphilus elongatus* M. Sars on its fan-worm host. *Proc. Zool. Soc. Lond.*, **133** 619.

—— (1962). Egg-number and ecology in commensal and parasitic copepods. *Ann. Mag. nat. Hist.* (13), **5**, 97.

Hand, C., and Hendrickson, J. R. (1950). A two-tentacled, commensal hydroid from California. *Biol. Bull.*, **99**, 74.

Hickok, J. F., and Davenport, D. (1957). Further studies in the behaviour of commensal polychaetes. *Ibid.*, **113**, 397.

Illg, P. L. (1958). North American copepods of the family *Notodelphyidae*. *Proc. U.S., Nat. Mus.*, **107**, 463.

Kawaguti, S. (1964a). Zooxanthellae in the coral are intercellular symbionts. *Proc. Jap. Acad.*, **40**, 7.

—— (1964b). An electron microscopic proof for a path of nutritive substances from zooxanthellae to the reef coral tissue. *Ibid.*, **40**, 10.

Limbaugh, C. (1961). Cleaning symbiosis. *Sci. Am.*, **205**, 42.

MacGinitie, G. E., and MacGinitie, N. (1949). *Natural History of Marine Animals*. McGraw-Hill, London.

Martin, R., and Brinckmann, A. (1963). Zum Brutparasitismus von *Phyllirhoë bucephala* PER. & LES. (GASTROPODA, NUDIBRANCHIA) auf der Meduse *Zanclea costata* GEGENB. (HYDROZOA, ANTHOMEDUSAE). *Pubbl. staz. zool. Napoli*, **33**, 206.

Mohr, J. L. (1958). Ciliates on crustaceans: The evolution of the Order Chonotrichida. *Proc. Internat. Congr. Zool. London,* **15,** 267.

Orton, J. H. (1921). The mode of feeding and sex phenomena in the pea-crab
(*Pinnotheres pisum*). *Nature,* **106,** 533.

Wickler, W. (1965). Mimicry and the evolution of animal communication. *Nature,*
**208,** 519.

Yonge, C. M., and Nicholls, A. G. (1931). Studies on the physiology of corals.
IV. The structure, distribution, and physiology of zooxanthellae. *Sci. Rept.
Gr. Barrier Reef Exped.,* **1,** 135.

*Index*

# Index